Plankton Communities

Edited by Leonel Pereira
and Ana Marta Gonçalves

Published in London, United Kingdom

IntechOpen

Supporting open minds since 2005

Plankton Communities
http://dx.doi.org/10.5772/intechopen.91092
Edited by Leonel Pereira and Ana Marta Gonçalves

Contributors
Guoqing Wang, John Moisan, Maria Emília Cunha, Hugo Quental Ferreira, Ana Barradas, Pedro Pousão Ferreira, Rocio Elizabeth Muciño Marquez, Dalila Aldana-Aranda, María Guadalupe Figueroa-Torres, Oscar Ubisha Hernández-Almeida, Evgeny Kurashov, Julia Krylova, Elena Protopopova, Leonel Pereira, Glacio Souza Araujo, Diana Pacheco, João Cotas, José William Alves da Silva, Jefferson Pablo de Souza Saboya, Renato Teixeira Moreira, Laura Mondragón-Camarillo, Salvador Rodríguez Zaragoza, Rosaura Mayén Estrada, Roberto Júnio Pedroso Dias, Mireya Ramírez-Ballesteros, Mariana Rossi, Carlos Alberto Durán-Ramírez, Gerardo Cruz-Jiménez, Margarita Reyes-Santos, Ana M. Marta Gonçalves, Fabiano Ramiro Serpe, Júlio César Rodrigues de Azevedo, Fernando J.M. Gonçalves, João C. Marques

First published in London, United Kingdom, 2022 by IntechOpen
IntechOpen is the global imprint of INTECHOPEN LIMITED, registered in England and Wales, registration number: 11086078, 5 Princes Gate Court, London, SW7 2QJ, United Kingdom
Printed in Croatia

British Library Cataloguing-in-Publication Data
A catalogue record for this book is available from the British Library

Additional hard and PDF copies can be obtained from orders@intechopen.com

Plankton Communities
Edited by Leonel Pereira and Ana Marta Gonçalves
p. cm.
Print ISBN 978-1-83968-608-5
Online ISBN 978-1-83968-609-2
eBook (PDF) ISBN 978-1-83968-610-8

For EU product safety concerns:
IN TECH d.o.o., Prolaz Marije Krucifikse Kozulić 3, 51000 Rijeka, Croatia, info@intechopen.com or visit our website at intechopen.com.

Meet the editors

Leonel Pereira has an undergraduate degree in Biology, a Ph.D. in Biology (specialty in Cell Biology), and a Habilitation degree in Biosciences (specialization in Biotechnology) from the Faculty of Science and Technology, University of Coimbra, Portugal, where he is currently a professor. In addition to teaching at this university, he is an integrated researcher at the Marine and Environmental Sciences Center (MARE), Portugal. His interests include marine biodiversity (algae), marine biotechnology (algae bioactive compounds), and marine ecology (environmental assessment). Since 2008, he has been the author and editor of the electronic publication MACOI – Portuguese Seaweeds Website (www.seaweeds.uc.pt). He is also a member of the editorial boards of several scientific journals. Dr. Pereira has edited or authored more than 20 books, 100 journal articles, and 45 book chapters. He has given more than 100 lectures and oral communications at various national and international scientific events. He is the coordinator of several national and international research projects. In 1998, he received the Francisco de Holanda Award (Honorable Mention) and, more recently, the Mar Rei D. Carlos award (18th edition). He is also a winner of the 2016 CHOICE Award for an outstanding academic title for his book *Edible Seaweeds of the World*. In 2020, Dr. Pereira received an Honorable Mention for the Impact of International Publications from the Web of Science

Ana Marta Gonçalves obtained a Ph.D. in Biology with a specialization in Ecology from the University of Coimbra, Portugal, in collaboration with Ghent University, Belgium, in 2011. Currently, she is an auxiliary researcher at the Marine and Environmental Sciences Center (MARE), Portugal, where she is also a member of the Directive Board. Since 2016, she has been a member of the Scientific Council of the Institute for Interdisciplinary Research, University of Coimbra (IIIUC). Dr. Gonçalves holds various administrative and management positions in international networks, societies (e.g., Society of Environmental Toxicology and Chemistry, AIL), and associations (e.g., PROAQUA). She is an editorial board member and reviewer for several indexed journals. She has published more than 70 journal articles, 50 book chapters, and 165 communications in international scientific events. She participated as a member and/or coordinator in more than twenty-five national and international projects and is currently the coordinator of four research projects. She has supervised more than ninety-five national and international undergraduate and graduate students. She has experience as a teacher of university courses and in accredited training sessions for teachers. Additionally, she has coordinated several ocean literacy and environmental education activities for kindergarten and school students. During her research career, Dr. Gonçalves obtained several grants and a MARS award for young scientists funded by The Royal Netherlands Institute for Sea Research (NIOZ). She has expertise in biosafety, biochemical pathways, and impacts of stressors in aquatic species. Her research focus is on the valorization of marine resources and their applications in the industrial sector, such as the food and pharmaceutical industries. Her studies also highlight the application of biomarker tools for monitoring and managing aquatic systems.

Contents

Preface

The term "plankton" was coined in 1884 by Hensen to indicate organisms randomly disposed in space. In 1974, plankton distribution was assumed by Lussenhop as uniformly distributed in space. Indeed, the term plankton means wandering; basically, it is a category of organisms that are passively displaced by water, that is, they are dragged by marine currents or even by waves. Many of the organisms that compose plankton have their own movements; however, because they belong to a small-scale water movement, they cannot overcome the force of currents or waves. Plankton is strongly influenced by climate features, chemical stressors, and hydrological conditions. Their sensitivity to these fluctuations results in communities that are continuously changing and adapting to environmental factors as well as to changes in available resources. The organisms that constitute plankton are subdivided into phytoplankton (autotrophic) and zooplankton (heterotrophic) in the marine food chain. Among the main planktonic organisms, we cite protozoa, microalgae, crustacean larvae, small crustaceans, jellyfish, and larvae of various animals. The marine plankton community can also be found in limnic (freshwater) environments. Within the planktonic community, we find zooplankton (animals), bacterioplankton (bacteria), and virioplankton (viruses), among other less significant groups. An interesting pattern is the vertical distribution of the copepod community, which is a complex structured response to tidal, diel, and lunar cycles, to environmental variables (e.g., salinity, temperature, hydrostatic pressures), and to reproductive cycles of prey and predators.

Colonies of unicellular algae with poor mobility form what is called phytoplankton (vegetables). Phytoplankton contains chlorophyll and other pigments, and like terrestrial plants, they can carry out photosynthesis. Due to these characteristics, they are essential for maintaining sea life, as they are the basis of the trophic chain of the aquatic universe. Phytoplankton organisms go beyond just being organic matter (food); they are also responsible for producing oxygen. In this production process, phytoplankton plants release oxygen gas in a dissolved form in water instead of releasing it directly into the atmosphere. The oxygen dissolved in the water is used by fish and other organisms that live in the sea to breathe.

Oceans are responsible for a large part of the oxygen that is produced on Earth and thus phytoplanktons are essential organisms for maintaining not only aquatic life but also life on the planet. Just like land vegetables, aquatic plants need light and nutrients to thrive. In this case, nutrients need to be dissolved in water; the main components of which are nitrogen, phosphorus, silicon, iron, and so on. Thus, the productivity of the oceans is associated with light conditions and the availability of these nutrients.

This book is divided into three sections. Section 1, "Introduction," includes two chapters: Chapter 1, "Response of Marine Plankton Communities in Ponds to the Presence of Vertical Structures" and Chapter 2, "Plankton: Environmental and Economic Importance for a Sustainable Future."

Section 2, "Phytoplankton," includes three chapters: Chapter 3, "Remote Sensing of Phytoplankton Pigments"; Chapter 4, "First Report on the Diversity of Epizoic Algae in Larval of Shellfish Gastropod *Aliger gigas*"; and Chapter 5, "The Use of Allelochemicals of Aquatic Macrophytes to Suppress the Development of Cyanobacterial "Blooms"."

Section 3, "Zooplankton," includes three chapters: Chapter 6 "Food Webs"; Chapter 7, "Ciliates as Symbionts"; and Chapter 8, "Changes in the Fatty Acids Profile of the Zooplankton Community Reveals the Quality of Four Reservoirs in the Hydroelectric Power Plants Located in the Iguaçu River, Paraná, Brazil."

Leonel Pereira
MARE—Marine and Environmental Sciences Centre,
Faculty of Sciences and Technology,
Department of Life Sciences,
University of Coimbra,
Coimbra, Portugal

Ana Marta Gonçalves
MARE—Marine and Environmental Sciences Centre,
Faculty of Sciences and Technology,
Department of Life Sciences,
University of Coimbra,
Coimbra, Portugal

Department of Biology and CESAM,
University of Aveiro,
Aveiro, Portugal

Acknowledgements

This work has the support of Fundação para a Ciência e Tecnologia, through the strategic projects UIDB/MAR/04292/2020 and UIDP/50017/2020+UIDB/50017/2020 granted to MARE and CESAM, respectively. Ana Marta Gonçalves acknowledges the University of Coimbra for the contract IT057 18-7253.

Section 1

Introduction

Chapter 1

Response of Marine Plankton Communities in Ponds to the Presence of Vertical Structures

Maria Emília Cunha, Hugo Quental Ferreira, Ana Barradas and Pedro Pousão-Ferreira

Abstract

The effects of bottom vertical structures like AquaMats® in enhancing plankton productivity was evaluated. One experimental earthen pond of 500 m^2 was provided with AquaMats® increasing the surface substrate area 12 times and water quality, phytoplankton and zooplankton populations developed during almost 100 days was compared with a pond without AquaMats®. Their presence favored the development of Dinoflagellates (Miozoa, Dinophyceae), mostly Gymnodiniales, which may be of some concern since some species of this group have been associated with toxic algal blooms while in the ponds without AquaMats® Diatoms (Bacillariophyta) predominate. In both ponds plankton production was very much sculptured by external nutrients added to the systems. The balance between different nutrients is extremely important to regulate the phytoplankton populations with Diatoms blooming at silicate concentrations higher than 2 μM and below this level and at low nitrate and high ammonium being more appropriate for Dinoflagellates. The linkage between phytoplankton and zooplankton population in ponds is strong with zooplankton exerting control over the phytoplankton population and vice-versa. The use of vertical substrates enhances plankton productivity by increasing the substrate area for periphyton fixation. The main zooplankton taxonomic groups associated with the presence of AquaMats® were Calanoid and Harpacticoid copepodites and nauplii, veligers of gastropods and trochophore of polychaets, larval stages of organisms that except for calanoid copepods are benthic and correspond to the meroplanktonic phase in the life cycle of those organisms.

Keywords: AquaMats®, phytoplankton, zooplankton, periphyton

1. Introduction

Decline in the world's fish stocks has led to an increasing demand for food from fish farming [1]. Much of this production is carried out in extensive and semi-extensive systems [2], mainly in Asia [1]. These systems are stocked with wild or farmed juveniles and rely on local natural productivity of lakes, earth ponds, reservoirs, and lagoons for feeding the fish and to maintain a good water quality and are characterized by low stocking densities and low to no inputs of food or fertilizers [2] and use of juveniles. Although intensification of these systems is a way to augment production, increasing profits are not likely to come from higher stocking densities due to the biological limits of these systems [3].

One alternative for productivity enhancement in these production systems is to use artificial substrates to enhance the colonization of the surface in ponds by periphyton [4]. These are complex mixtures of algae, cyanobacteria, heterotrophic microbes, and detritus that are attached to submerged surfaces and are the primary producers in streams, providing food for benthic invertebrates, which feed fish and other invertebrates [4]. Many of these organisms possess life cycles with meroplanktonic phases that boost zooplankton abundance. This increase of plankton abundance can be used to advantage for rearing fish in the ponds since it provides food for their first larval stages and therefore adds to the profitability of such systems. Besides the saving in cost of fry, juveniles produced in these natural systems are better adapted to grow out conditions in ponds. Another benefit is the possibility of using such systems for species diversification since the small prey produced will enable the larviculture of some marine fish species with small mouth gapes such as groupers [5].

Types of artificial substrates used for periphyton-based aquaculture are mainly natural substrates such as tree branches used in some African countries and mangrove leaves and twigs used in Asia [6]. Also, in Asia, bamboo has been intensively studied and already incorporated successfully in farms with an established protocol. Pilot studies in periphyton have also been performed using plastic mesh sheets and nets [6].

AquaMats® are another artificial substrate used in aquaculture trials [7–9] and are widely used for advanced natural biofiltration in ponds/lakes. They are flexible curtains of highly specialized synthetic substrates used to increase the vertical surfaces of lagoons or ponds. Each curtain provides a three-dimensional surface with approximately 200 m^2 of effective surface area which is a benefit for fixation of live organism in a flat two-dimensional surface. The increase of pond surface area by the presence of vertical substrates leads to a larger colonization area for sessile biota that attach to the substratum. This biota will contribute to the enhancement of primary and secondary productivity (mainly benthic but also pelagic) that in addition to their larvae will increase the feed abundance for fish larvae. The present work presents the results of a trial to evaluate the effect of AquaMats® on the plankton species composition and productivity and water quality in earthen ponds.

2. Material and methods

Facilities of the Aquaculture Research Station (37° 02' N; 07 ° 49' W), of the Portuguese Institute for the Sea and Atmosphere (IPMA - for Instituto Português do Mar e Atmosfera), based in Olhão, southern Portugal, were used for the trial (**Figure 1**).

2.1 Experimental setting

Two rectangular earthen ponds of 750 m^3 each (1.4 m mean water depth) were used to study the effect of the presence of vertical substrates on the species composition and abundance of the phyto and zoo plankton populations. Before the experiment, the floor of the earthen ponds was thoroughly washed to remove organic sediment and dried for two weeks to allow better oxygenation of the anaerobic layers by direct exposure to air and sunlight [10]. After this period 30 bottom deployment format (BDF) AquaMats® were set up in one of the earthen ponds arranged in 10 rows perpendicular to the water flow. Each AquaMat® had an

Figure 1.
Geographical location of the Aquaculture Research Station (EPPO) of the Portuguese Institute for the Sea and Atmosphere and of the experimental ponds in blue and orange lines. Blue: Pond with AquaMats®; Orange: Control pond (without AquaMats®).

effective surface area of 208 m^2 (www.AquaMats.com) and their presence increased by 9 times the effective surface area of the pond (681 m^2). The second earthen pond remained without AquaMats® as a control. Earthen ponds were filled on May 9 with sand-filtered seawater from the Ria Formosa coastal lagoon (Algarve, Portugal). Additional water from adjacent ponds used for fish on-growing was pumped into each earthen pond as a fertilizer to boost initial plankton production. After filling, water exchange was set at 10% renovation day^{-1} during the entire trial. No aeration was provided during the experiment. Organic fertilizer (alfalfa pellets) at 28 g m^{-2} [11] were uniformly distributed in the two earth ponds two weeks after filling (May 22) and every 10 days thereafter. The trial ended on August 15.

2.2 Water quality and plankton monitoring

Temperature, salinity, dissolved oxygen, and pH were measured daily using a portable meter (HI9828 - Hanna Instruments®). Monitoring of major inorganic nutrients analysis (total ammonia nitrogen (TAN), nitrate and nitrite (NO_3-N and NO_2-N), orthophosphate (PO_4-P) and silica (SiO_2), was also performed daily during the first week after fertilization, every other day in the second week and weekly thereafter, as well as solid particulate matter (SPM), chlorophyl *a* and identification and enumeration of phytoplankton and zooplankton populations. Water samples of 10 L were collected near the water inlet, in the middle and at the outlet, pooled together. One liter of water was used for analysis of nutrients, half litter for chlorophyll *a* estimate, another half litter for phytoplankton analysis and the remaining water (28 L) filtered throughout a 55 μm plankton mesh for zooplankton counts.

Phytoplankton samples were preserved with a few drops of Lugol's iodine and zooplankton in 4% buffered formaldehyde.

2.3 Laboratorial analysis

Inorganic nutrients were determined by colorimetry [12] using a "Skalar" auto-analyzer with a detection limit of 0.2 μM for ammonium and 0.05 μM for nitrite, nitrate, phosphate, and silicate. Chlorophyll α was determined by spectrophotometry after passing the water sample through 0.47 μm cellulose nitrate membrane filters (Type 11306, Sartorius Stedim Biotech) and extracted using 10 ml acetone. Calculation was done using the formula from [13]:

$$\text{Chl}\,a\left(\mu g / L\right) = 11.0 \times 2.43 \times \left(665 - 665_a\right) \times 10 / \left(V_f\right) \quad (1)$$

where, 665 is the absorbance before acidification, 665_a is the absorbance after acidification, and V_f is the amount filtered water (liters).

Phytoplankton enumeration and identification was done under an inverted microscope, after sedimentation during 24 h of 50 mL sub-sample. Zooplankton present in 28 L water samples were identified and enumerated under a stereomicroscope.

2.4 Data analyses

Two diversity indices were calculated for each plankton sample: Taxa Richness and Margalef Diversity index. The taxa richness (T) was the number of taxa present in the sample while the Margalef diversity index, $d = (T - 1)/\ln N$, was the number of taxa (T), weighted by N, the total number of individuals in the sample.

Data were analyzed for normality and ANOVA tests were used for comparison between means and the rejection level for the null hypothesis was $P = 0.05$. Comparison of means were based on log (x + 1) transformed data, but values depicted here are not transformed. Regression analysis were used to assess the significance of the relationship between time and the outcome variable.

3. Results and discussion

3.1 Environmental conditions

Temperature during the trial varied between 21.3°C and 26.5°C and salinity between 33.3 and 37.3 (**Figure 2**). The Control pond presented slightly higher temperatures after the first month of trial but neither of these parameters presented significant mean differences (**Table 1**). Both, Dissolved Oxygen (DO) and pH were significantly lower in the Control pond with a more noticeable descend 42 days after the beginning of the trial when salinity also increased.

Nutrient concentrations were below autoanalyzer detectable levels during the first three weeks of the trial but raised after the first addition of alfalfa, on the 22nd day after filling the ponds (**Figure 3**). In general, they showed spikes of increase responding to preceding fertilizations. Mean concentration of HPO_4^{2-} SiO^2, NH^{4+} and NO^{3-} were not significantly different between ponds (**Table 1**). Although DO and pH were significantly higher in the ponds with AquaMats®, suggesting higher primary production, chlorophyll *a* concentration was significantly lower (**Table 1**).

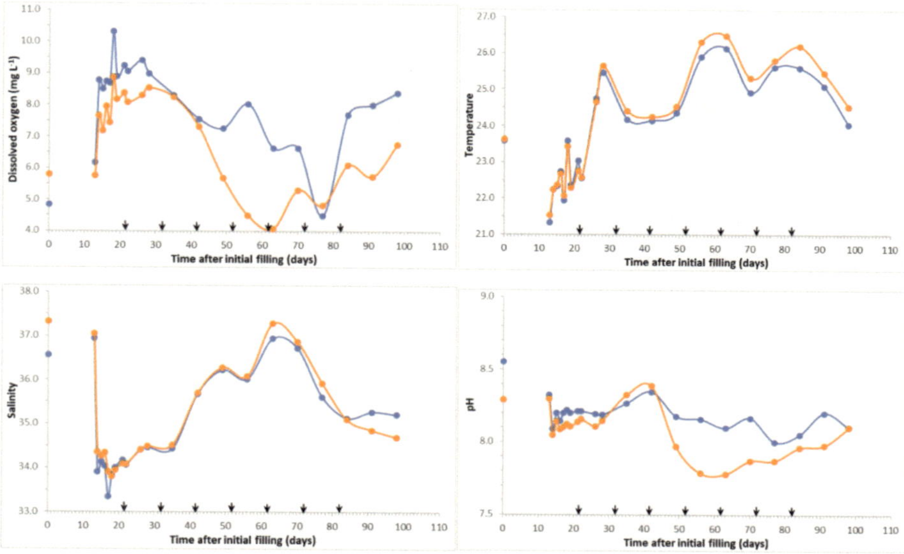

Figure 2.
Temporal evolution of temperature, salinity, dissolved oxygen, and pH in the ponds during the trial. Black arrows designate fertilization of the ponds. Orange dots and lines: Control pond; blue dots and line: Ponds with AquaMats®.

Parameters	N	Control pond	AquaMats®
Temp (°C)	23	24.1 ± (1.6)	23.9 ± (1.4)
Salinity	23	35.2 ± (1.2)	35.0 ± (1.1)
DO (mg L^{-1})	23	6.8 ± (1.5) *	7.9 ± (1.4) *
pH	23	8.1 ± (0.2) **	8.2 ± (0.1) **
NH_4 (μM)	23	0.43 ± (0.60)	0.40 ± (0.62)
$NO_3 + NO_2$ (μM)	23	0.27 ± (0.32)	0.24 ± (0.44)
$Si(OH)_4$ (μM)	23	2.29 ± (2.46)	1.77 ± (2.68)
HPO_4^{2-} (μM)	23	0.13 ± (0.13)	0.14 ± (0.19)
Chl_a (μg L^{-1})	23	3.68 ± (1.88) *	2.60 ± (1.24) *
SPM (mg L^{-1})	23	68.0 ± (10.8)	70.6 ± (12.8)

**P > 0.05.*
***P > 0.01.*
SPM – Suspended particulate matter.

Table 1.
Mean concentration ± (S.D.) of the referred environmental paraments inside the ponds.

The trial started with similar concentration of chlorophyll *a* in both ponds and the concentrations rose after a short period of acclimatation (**Figure 3**). The increase in chlorophyll *a* due to the uptake of nutrients by the phytoplankton lead to a complete depletion of nutrients and to a consequent drop on the chlorophyll *a*. After the first alfalfa fertilization, that occurred on the 22nd day, there was an increase of Chl_*a* concentration in both ponds. In general, the addition of nutrients to the earthen ponds produce effect on the increase of the chlorophyll *a* concentration since it contributed to increase the availability of nitrogen, phosphorus, and silicate in the ponds (**Figure 3**). Solid Particulate Matter (SPM) concentrations, which included

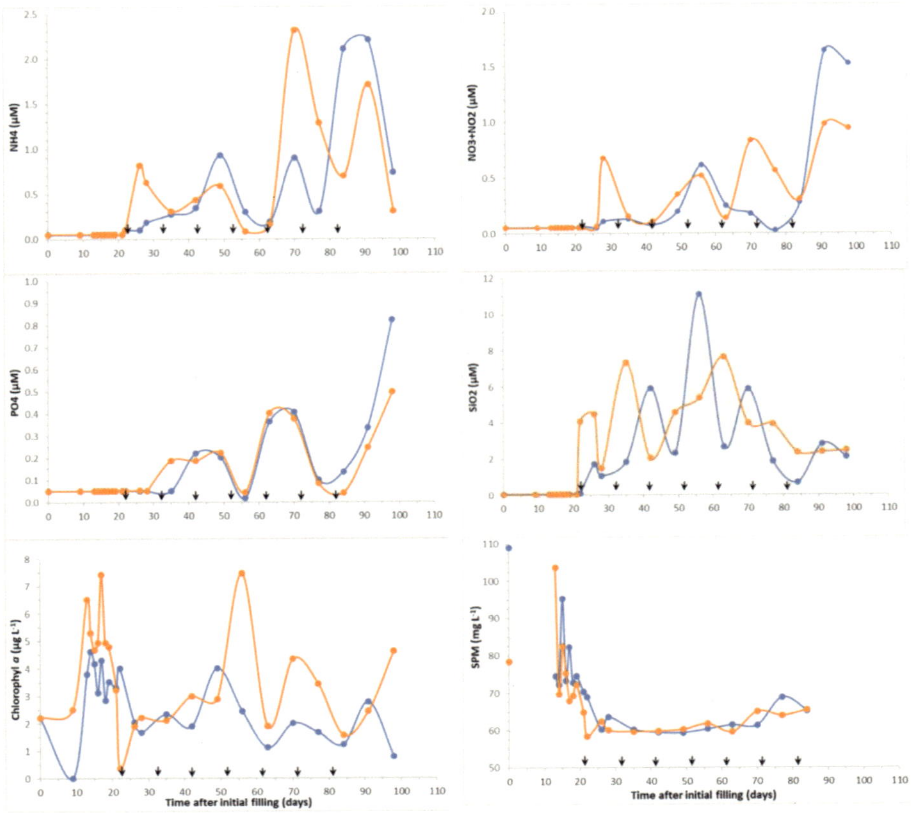

Figure 3.
Temporal evolution of ammonia, nitrate+nitrite, phosphate, silicate, chlorophyll a concentration and solid particulate matter (SPM) in the ponds during the trial. Black arrows designate fertilization of the ponds. Orange dots and lines: Control pond; blue dots and line: Ponds with AquaMats®.

plankton cells, were not significantly different among ponds. Concentrations were high at the start of the trial and decrease steadily during the first three weeks after which they increased slowly until the end of the trial.

3.2 Phytoplankton

Phytoplankton species and mean abundances in the ponds during the trial are presented in Appendix A. Phytoplankton densities in the Control pond and in the pond with AquaMats® showed similar patterns of evolution with four blooms followed by crashes during the monitored period (**Figure 4**). Initial cell abundances in the water of the two ponds were high followed by a sharp decrease in the phytoplankton. Recovery occurred after the first two weeks with a bloom that lasted for a week followed by a crash. After the initial fertilization, high concentrations were reached and phytoplanktonic abundance seemed to be sustained despite some decreases. Although, during the first month of the trial mean phytoplankton densities tended to be lower in the Control pond, mean concentrations for the entire trial were not significantly different between the two ponds (**Table 2**). The pond with AquaMats® have sharper variations in phytoplankton abundances and the Control pond had more steady densities that increased with time. Diatoms and dinoflagellates composed the bulk of the phytoplankton population in both ponds but non-identified phytoflagellates were significantly more important in the pond with Aquamats®.

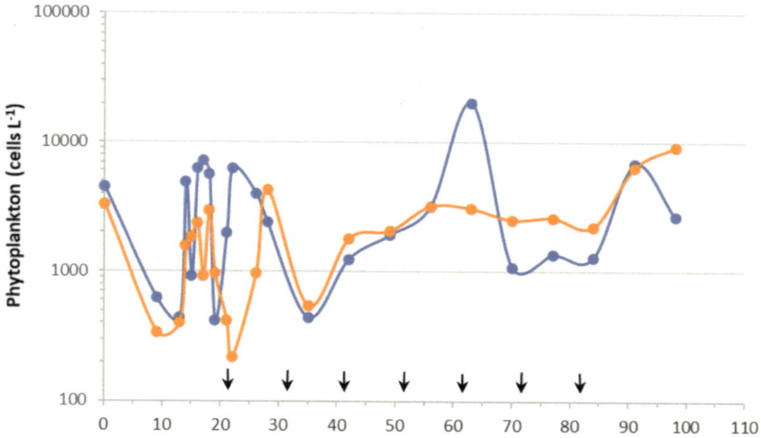

Figure 4.
Temporal progression of phytoplankton abundance in ponds. Orange dots and lines: Control pond; blue dots and line: Ponds with AquaMats®; black arrows designate fertilization of the ponds.

	N	Control	With AquaMats
Total phytoplankton (# L^{-1})	23	2,340 ± (2,082)	3,557 ± (4,174)
Taxa richness	23	9.3 ± (2.2)	9.5 ± (2.3)
Margaleff index	23	1.91 ± (0.40)	1.84 ± (0.42)
Diatoms (# L^{-1})	23	1,390 ± (1,242)	2,066 ± (4,073)
Dinoflagellates (# L^{-1})	23	693 ± (1,513)	951 ± (975)
Phytoflagellates n.i. (# L^{-1})	23	47 ± (90) *	465 ± (934) *
Others (# L^{-1})	23	210 ± (391)	75 ± (6)

*$P > 0.05$.

Table 2.
Phytoplankton estimates and abundance of main groups in ponds.

Taxa Richness and the Margaleff Index were not significantly different between ponds with dominance of diatoms over dinoflagellates in both (**Table 2**). In general phytoplankton abundance was higher in the pond with AquaMats® but the difference was not statistically different from the Control (**Table 2**). Both ponds recorded taxa richness minima immediately after the beginning of the experiment followed by maxima two to three weeks after filling. Species richness remained relatively leveled afterwards (**Figure 5**). Regressions between taxa richness and time after filling, shown in the graphs, were not significant suggesting that the number of taxa present in the ponds were independent of the time.

The temporal succession of phytoplankton groups in the Control pond was essentially dominated by diatoms with the phytoplankton blooms preceded by silicate maxima (**Figure 3**). Exceptions were the first two weeks of the experiment and for two consecutive samples on days 56th and 63rd and again on day 100th after filling, when dinoflagellates of the genus *Gymnodinium* became the most abundant group. In this pond during this first month *Navicula* spp. predominate and after the first month of trial Diatoms were mostly *Cylindrotheca closterium*, (**Figure 5** and **Figure 6** – upper graphs). Comparatively, the pond with AquaMats® started, during the first two weeks, with higher relative abundance of diatoms, mostly *C. closterium*, followed by a sustained period of more than one month with higher

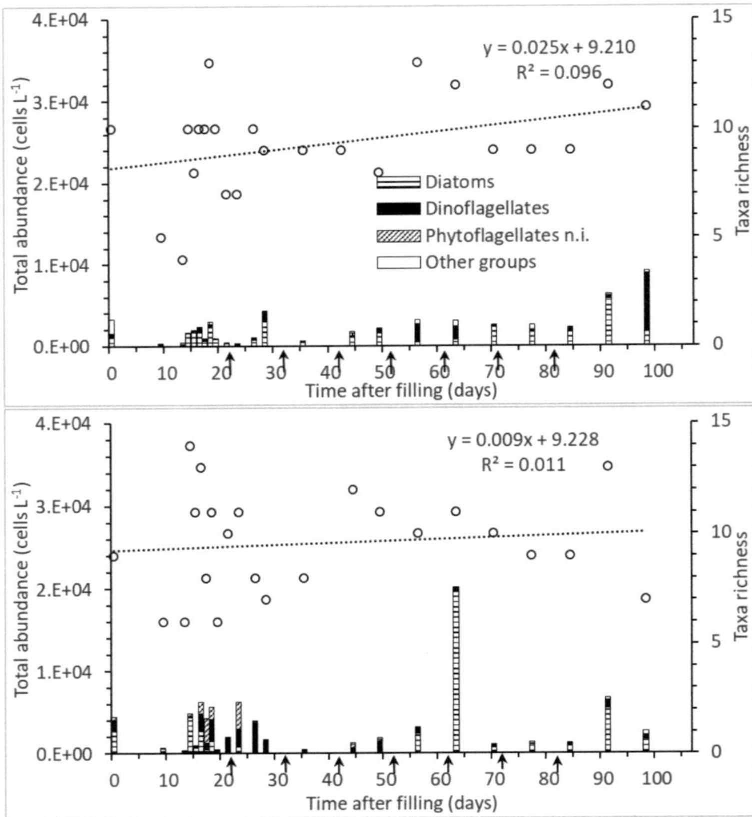

Figure 5.
Temporal evolution of the abundance of main phytoplankton groups and taxa richness. Upper graph – Control pond; lower graph – Pond with AquaMats®. Black arrows designate fertilization of the ponds.

concentrations of dinoflagellates although non identified phytoflagellates were also important (**Figure 5** and **Figure 6** – lower graphs). *C. closterium* and other Pennate diatoms, became more important in the second half with exception for the last sampling day when non-identified dinoflagellates dominate. The dinoflagellate community, initially dominated by *Prorocentrum micans*, was replaced by individuals of the genus *Gymnodinium* immediately after the first alfalfa fertilization dominating in the samples around the 30th day after filling. The highest abundances of dinoflagellates were observed at the end of the experiment in both ponds suggesting a seasonal effect.

3.3 Zooplankton

Both the mean abundance of zooplankton individuals and the taxa richness (T) were not significantly different among ponds but the Margaleff index was significantly higher in the pond with AquaMats® due to lower zooplankton abundance while the number of species remained similar (**Table 3**).

The temporal progression of zooplankton abundance was similar among the two experimental ponds and showed that after filling there was a decrease followed by a week of adjustment when biomass was low (**Figure 7**). This adjustment period ended with the progressive increase in the number of zooplankton organisms and, similarly to what happened to the phytoplankton temporal evolution, there were four peaks of higher abundance followed by crashes. The periods of higher

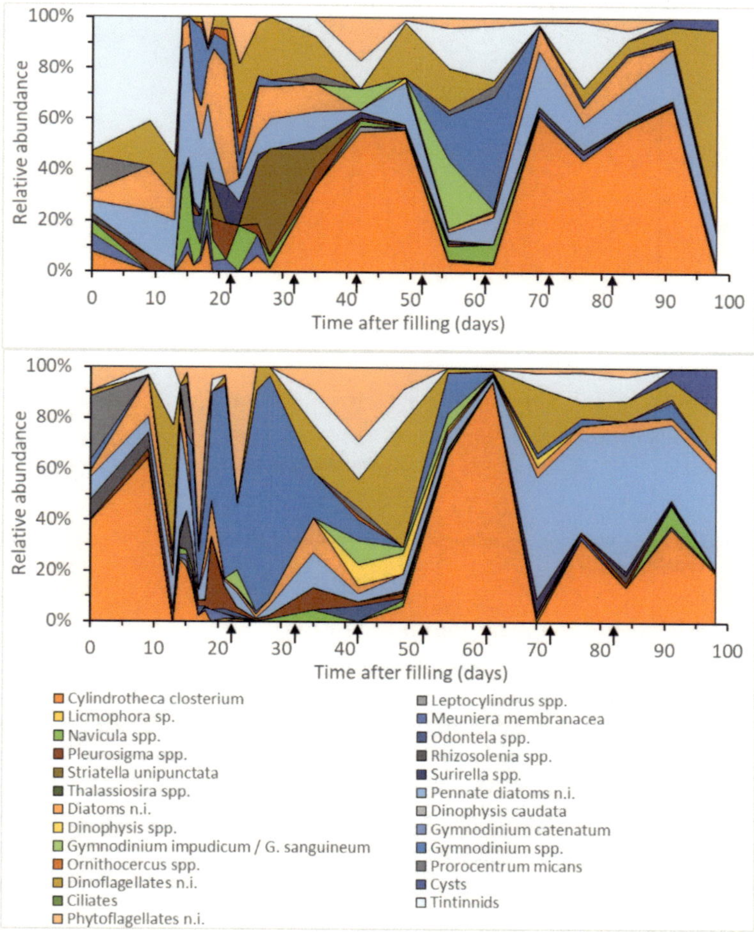

Figure 6.
Temporal succession of phytoplankton populations. Black arrows designate fertilization of the ponds. Upper graph – Control pond; lower graph – Pond with AquaMats®.

	Control	AquaMats®
Total zooplankton (# L^{-1})	38.5 ± (27.3)	33.8 ± (41.1)
Margaleff index	1.7 ± (0.5) *	2.2 ± (1.0) *
Taxa richness (T)	6.6 ± (1.0)	6.5 ± (1.2)
$^*P > 0.05$.		

Table 3.
Mean zooplankton estimates in ponds.

zooplankton abundance occurred days after the phytoplankton blooms suggesting a strong zooplankton control over the phytoplankton population. Although zooplankton abundance presented similar patterns of development of booms and crashes in the ponds, the abundance during the first half of the trial (45 days) was significantly higher in the Control pond (**Figure 7**). The taxa richness started also to be higher in the Control pond but remained relatively constant over time in this pond as suggested by the regression equation in **Figure 8** while in the pond with AquaMats® it was lower at the beginning of the trial but increased significantly with time.

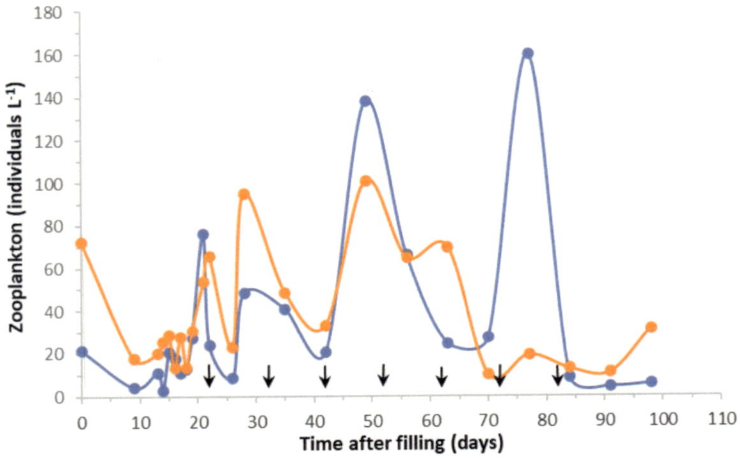

Figure 7.
Temporal progression of zooplankton abundance in ponds. Orange dots and lines: Control pond; blue dots and line: Ponds with AquaMats®.

Zooplankton species in the ponds and their mean abundances are listed in Appendix B. From the most frequent zooplankton groups, veligers (both Gastropoda and Bivalvia), adults from Calanoida copepods and Cyclopoida naplii were significantly more abundant in the Control pond. With exception of Polychaeta larvae, Calanoida nauplii, Harpacticoida copepodids and Cirripedia nauplii, were most abundant in the pond provided with AquaMats® although differences were not significant. The Calanoid copepod *Acartia clausi* was only present in the Control pond while *Paracartia grani* was mostly present in the pond with AquaMats®. Although not significantly different, their nauplii were more abundant in the pond with Aquamats®.

In both ponds there was a fairly number of Calanoida nauplii and Polychaeta larvae. Nine days after, Calanoid adults (*Acartia clausi*) became more important in the Control pond remaining the most important taxa during the following week (**Figure 8**, upper graph). The first boom, on day 21, was mostly composed by those adults and nauplii. The following booms were mostly formed by copepod nauplii (both Calanoida, Cyclopoida and Harpaticoida). Gastropoda veligers, that were always present, became more important by the end of the experiment. In the pond with AquaMats®, Calanoid copepod nauplii were also important at start of the experiment but in the following sampling period Polychaeta larvae became gradually more abundant and continued doing so for the following month (**Figure 8**, lower graph). By then Harpacticoid copepod adults became evident. Succeeding booms were composed mainly by copepod nauplii (mostly Calanoida but also Harpaticoida).

4. Discussion

There were no significant differences in phytoplankton and zooplankton densities among the two ponds, but parameters related to plant production, such as dissolved oxygen (DO) and pH, showed significantly higher values in the ponds with AquaMats® suggesting higher primary production in this pond. The combination of these two parameters with lower nitrate, ammonia, silicate and in the pond with AquaMats® further suggests greater overall algal production in this treatment, which was not reflected in the Chl_*a* concentration (**Table 2**). Therefore, the higher

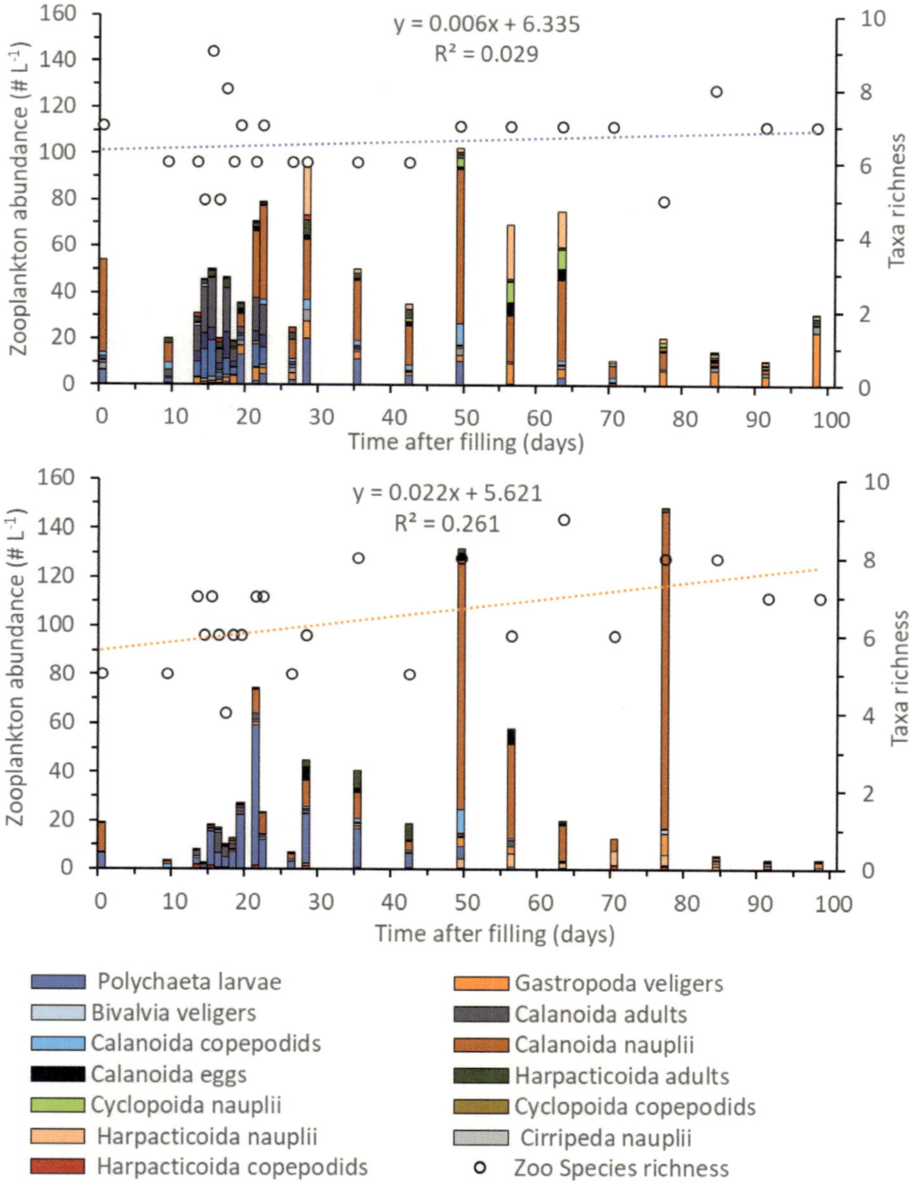

Figure 8.
Temporal evolution of the abundance of main zooplankton groups and species richness. Upper graph – Control pond; lower graph – Pond with AquaMats®. Black arrows designate fertilization of the ponds.

primary production was probably associated with the periphyton developed in the AquaMats® and to a lesser extent to the slightly higher, although no significant, phytoplankton population.

The presence of diatoms, dinoflagellates and non-identified phytoflagellates are common in fish and oyster integrated production in earthen ponds that supplied the dissolved nutrients required by the phytoplankton [14]. In the present case, there were no fish and oyster production, but external nutrients were supplied by alfalfa. The temporal fluctuations in abundance of phytoplankton were very much connected to the regular supply of alfalfa with a strong increase immediately after fertilization. In general the phytoplankton blooms followed silicate maxima and they were dominate by diatoms, mostly

Cylindrotheca closterium, although Pennate diatoms started to be more important in the Control pond. The success of the diatom group seemed to be due to a high inherent growth rate at non-limiting silicate concentrations [15]. However Dinoflagellates and Phytoflagellates also marked their presence in the pond with AquaMats® when fertilization started and dominate for almost a month. This was a period of still low nitrate and silicate concentrations and relative higher rates of $NH4^+$ in the pond with AquaMats® which may be a possible explanation for the higher number of dinoflagellates in this pond [16]. Among the dinoflagellates present in the pond with AquaMats®, Gymnodiniales were the most important group.

Zooplankton abundance presented similar patterns of development of booms and crashes in the ponds and occurred days after the phytoplankton bloom suggesting a strong zooplankton control over the phytoplankton population. The abundance during the first half of the trial was significantly higher in the Control pond and the taxa richness was also higher remaining relatively constant over time. Calanoida (*Acartia clausi*) adults and nauplii and Polychaeta larvae composed mostly of the population during this time. In the pond with AquaMats®, zooplankton abundance and taxa richness were both initially lower and increased significantly over time reflecting the effect of the disturbance caused by the deployment of the AquaMats® in the ponds and the consequent recovery. Polychaeta larvae, abundant during the 45 days, were overrun mostly by Calanoida nauplii, and to a lesser extent by Harpacticoida nauplii and Gastropoda veligers. These are larval stages of organisms that except for calanoid copepods are benthic. At the example of *Acartia clausi* the adult calanoid present in the ponds with AquaMats® (*Paracartia grani*) reproduce by shedding eggs that attach to substrates [17–19]. These eggs can be subitaneous or diapause but in both cases, they need light to hatch [20]. The presence of AquaMats® as vertical substrates leads to an increase in the areas where the eggs can be attached and where they remain exposed to light and ready to hatch, may explain the higher number of Calanoid nauplii.

5. Conclusions

Plankton production in ponds is very much sculptured by external nutrients added to the systems and therefore fertilization and maintaining the balance between different nutrients is extremely important to control the phytoplankton populations. The linkage between phytoplankton and zooplankton population in ponds is strong with zooplankton exerting control over the phytoplankton population and vice-versa.

The use of vertical substrates like AquaMats® seemed to be able to enhance plankton productivity by increasing the substrate area for periphyton fixation. Their presence favored the development of Dinoflagellates, mostly Gymnodiniales, which may be of some concern since some species of this group have been associated with toxic algal blooms. The main zooplankton taxonomic groups associated with the presence of AquaMats® were Calanoid and Harpacticoid copepodids and nauplii, veligers of gastropods and trocophora of polychaets. These are all small larval stages of organisms that are important as food for fish larvae during the first phases of development and therefore there is potential for the use of AquaMats® in mesocosms for rearing fish larvae in semi-intensive systems either for the quality of the farmed juveniles or to rear species with larval stages that only survive with natural food increasing aquaculture diversification.

Acknowledgements

This work was carried out under the INTERREG V A Espanha Portugal (POCTEP) program, project 0750_AQUA_AMBI_2_5_P "AQUA&AMBI 2 - Apoio à gestão das zonas húmidas do litoral do Sudoeste Ibérico: interações entre a Aquacultura e meio Ambiente na região transfronteiriça Alentejo-Algarve-Andaluzia - FASE 2".

Appendices and nomenclature

Appendix A. Mean abundance of phytoplanktonic taxa during the trial (cells L^{-1})

Planktonic species	N	Control	Aquamats
BACILLARIOPHYCEAE			
Cylindrotheca closterium	23	515 ± (934)	1,296 ± (3,940)
Leptocylindrus spp.	23	7 ± (16)	8 ± (20)
Licmophora sp.	23	1 ± (4)	4 ± (8)
Meuniera membranacea	23	41 ± (66)	17 ± (35)
Navicula spp.	23	114 ± (161)	48 ± (121)
Odontela spp.	23	10 ± (29)	18 ± (41)
Pleurosigma spp.	23	33 ± (33)	43 ± (54)
Rhizosolenia spp.	23	7 ± (20) *	60 ± (142) *
Striatella unipunctata	23	97 ± (362) *	1 ± (4) *
Surirella spp.	23	17 ± (21) **	7 ± (18) **
Thalassiosira spp.	23	0 ± (0)	8 ± (38)
Pennate diatoms n.i.	23	389 ± (381)	457 ± (578)
Diatoms n.i.	23	160 ± (142)	98 ± (124)
DINOPHYCEAE			
Dinophysis caudata	23	1 ± (4)	2 ± (6)
Dinophysis spp.	23	3 ± (9)	15 ± (39)
Gymnodinium catenatum	23	0 ± (0)	7 ± (26)
Gymnodinium impudicum/ G. sanguineum	23	44 ± (180)	35 ± (99)
Gymnodinium spp.	23	151 ± (307) *	577 ± (859) *
Ornithocercus spp.	23	6 ± (9)	5 ± (11)
Prorocentrum micans	23	33 ± (95)	137 ± (347)
Dinoflagellates n.i.	23	455 ± (1,472)	173 ± (219)
OTHER			
Cysts	23	24 ± (86)	32 ± (110)
Ciliates	23	0 ± (0)	1 ± (4)
Tintinnids	23	185 ± (394)	42 ± (70)
Phytoflagellates n.i.	23	47 ± (90) *	465 ± (934) *

*$P > 0.05$.
**$P > 0.01$.

Appendix B. Mean abundance of zooplankton taxa during the trial

	N	Control	AquaMats
		(# L-1)	(# L-1)
Polychaeta (larvae)	23	3.6 ± (5.3)	9.0 ± (13.0)
Gastropoda (veliger)	23	4.2 ± (4.8) *	1.7 ± (3.2) *
Bivalvia (veliger)	23	1.4 ± (1.1) **	0.6 ± (0.7) **
Acartia clausi (female)	23	3.0 ± (4.5) **	0.0 ± (0.0) **
Paracartia grani (female)	23	0.2 ± (0.3) **	0.9 ± (1.1) **
Acartia spp. (male)	23	2.0 ± (3.1) *	0.6 ± (0.8) *
Acartia spp. (copepodite)	23	1.4 ± (2.1)	0.8 ± (2.0)
Calanoida spp. (nauplii)	23	14.7 ± (17.8)	15.5 ± (33.2)
Calanoida spp. (egg)	23	1.0 ± (1.4)	0.8 ± (1.6)
Oithona spp. (copepodite)	23	0.1 ± (0.3)	0.0 ± (0.1)
Oithona spp. (nauplii)	23	1.2 ± (2.5) *	0.0 ± (0.1) *
Harpaticoida spp. (adults)	23	1.9 ± (1.6)	1.1 ± (1.9)
Harpaticoida spp. (copepodite)	23	0.5 ± (0.6)	0.6 ± (0.7)
Harpaticoida spp. (nauplii)	23	3.1 ± (7.0)	1.6 ± (2.4)
Cirripeda spp. (cypris)	23	0.9 ± (4.1)	0.1 ± (0.3)
Cirripeda spp. (nauplii)	23	0.1 ± (0.2)	0.2 ± (0.4)

*$P > 0.05$.
**$P > 0.01$.

Author details

Maria Emília Cunha[1*], Hugo Quental Ferreira[1], Ana Barradas[2] and Pedro Pousão-Ferreira[1]

1 Olhão Aquaculture Research Station, Av. do Parque Natural da Ria Formosa, Olhão, Portugal

2 Biosystems and Integrative Sciences Institute (BioISI), Faculdade de Ciências, Universidade de Lisboa, Lisboa, Portugal

*Address all correspondence to: micunha@ipma.pt

References

[1] FAO, 2020. The State of World Fisheries and Aquaculture 2020. Sustainability in action. Rome. DOI/10.4060/ca9229en.

[2] FAO, 2000. Small ponds make a big difference. Integrating fish with crop and livestock farming. Rome, FAO. 2000. 30p.

[3] Wurts, W. A., 2000. Sustainable Aquaculture in the Twenty-first Century. Reviews in Fisheries Science 8 (2), 141-150.

[4] Azim, M.E., Verdegem, M.C.J., van Dam, A.A., Beveridge, M.C.M., 2005. Periphyton: Ecology, Exploitation and Management. CABI Publishing. U.K.

[5] Shields, R. J., 2001. Larviculture of marine finfish in Europe. Aquaculture 200, 55-88.

[6] Van Dam, A.A., Beveridge, M.C.M., Azim, M.E., Verdegem, M.C.J., 2002. The potential of fish production based on periphyton. Reviews of Fish Biology and Fisheries 12, 1-31.

[7] Bratvold, D., Browdy, C.L., 2001. Effect of sand sediment and vertical surfaces (AquaMats®) on production, water quality and microbial ecology in an intensive *Litopenaueus vannamei* culture system. Aquaculture 195, 81-94.

[8] Moss, K.R.K., Moss, S.M., 2004. Effects of artificial substrate and stocking density on the nursery production of Pacific white shrimp *Litopenaueus vannamei*. Journal of World Aquaculture Society 35, 536-542.

[9] Otoshi, C.A., Montgomery, A.D., Matsuda, E.M., Moss, S.M., 2006. Effects of artificial substrate and water source on growth of juvenile pacific white shrimp, Litopenaueus vannamei. 37, 210-213.

[10] Boyd, C.E., Tucker, C.S., 1998. Pond Aquaculture Water Quality Management. Kluwer Academic Publishers. Netherlands.

[11] Ludwig, G.M., Stone, N.M. and Collins, C., 1998. Fertilization of fish fry ponds. Southern Region Aquaculture Center. Publication N° 469.

[12] Hansen, H.P., Koroleff, F. Determination of nutrients. In: Grashoff, K., Kremling, K., Ehrhardt, M., editors, Methods of Seawater Analysis. 1999, 159-228. DOI/10.1002/9783527613984.ch10

[13] Lorenzen, C.J., 1967. Determination of Chlorophyll and Pheopigments Spectrophotometric Equations. Limnology and Oceanography, 12, 343-346.

[14] Cunha, M.E., Quental-Ferreira, H., Parejo, A., Gamito, S., Ribeiro, L., Moreira, M., Monteiro, I., Soares, F., Pousão-Ferreira, P., 2019. Understanding the individual role of fish, oyster, phytoplankton and macroalgae in the ecology of integrated production in earthen ponds. Aquaculture 512, 734297

[15] Egge, J. K. Aksnes, D. L., 1992. Silicate as regulating nutrient in phytoplankton competition. Marine Ecology Progress Series 83, 281-289

[16] Rees, T.A.V., 2007. Metabolic and ecological constraints imposed by similar rates of ammonium and nitrate uptake per unit area at low substrate concentrations in marine phytoplankton and macroalgae. Journal of Phycology 43. 197-207.

[17] Uye, S., Fleminger, A., 1976. Effects of various environmental factors on egg development of several species of *Acartia* in Southern California. Marine Biology 38, 253-262.

[18] Lindley, J.A., 1997. Eggs and their incubation as factors in the ecology of planktonic crustacean. Journal of Crustacean Biology 17, 569-576.

[19] Dumont, H.J., Nandini, S., Sarma, S.S.S., 2002. Cyst ornamentation in aquatic invertebrates: a defense against egg-predation. Hydrobiologia 486, 161-167.

[20] Peck, M.A., Ewest, B., Holste, L., Kanstinger, P., Martin, M., 2008. Impacts of light regime on egg harvests and 48-h egg hatching success of *Acartia tonsa* (Copepoda: Calanoida) within intensive culture. Aquaculture 275, 102-107.

Chapter 2

Plankton: Environmental and Economic Importance for a Sustainable Future

Glacio Souza Araujo, Diana Pacheco, João Cotas, José William Alves da Silva, Jefferson Saboya, Renato Teixeira Moreira and Leonel Pereira

Abstract

Plankton is composed by unicellular, filamentous or colonial organisms that may have prokaryotic or eukaryotic cell structures. These organisms have an extreme ecological importance in the different water bodies worldwide, as they fix carbon dioxide, produce oxygen and are an important key element in the basis of various food chains. Through an industrial perspective, phytoplankton species have been used as a feedstock for a wide range of applications, such as wastewater treatment, or production of high value compounds; and commercial products, such as food and feed supplements, pharmacological compounds, lipids, enzymes, biomass, polymers, toxins, pigments. Zooplankton is commonly used as live food for larval stages to the period of termination of fish, shrimp, mollusks and corals. These types of organisms have characteristics such as a valuable nutritional composition, digestibility, buoyancy, ease of ingestion and attractive movement for post-larvae, thus presenting economic importance. This book chapter aims to demonstrate the several advantages that plankton have, their ecological and economic importance, targeting the production of add-value products.

Keywords: phytoplankton, zooplankton, bioactive compounds, industrial products

1. Introduction

Oceans cover 71% of the surface of the Earth and have a huge diversity and high percentage of the earth biota [1]. Oceans take a key role in the global carbon cycle, therefore openly influence the speed and magnitude of climate changes, which can be observed in the aquatic organisms [2]. Moreover, the biota of the oceans have huge socioeconomic value, through food and feed production, nutrient recycling and carbon dioxide regulation [3]. Climate changes impacts on the ocean biota will provoke economic implications, so there is a need to understand the key drivers to understand the ecological change and how some to exploit the ocean organisms without putting pressure in the surrounding ecosystem [4]. In which, phytoplanktonic microorganisms develop the basis to the food chain status quo and greatly contribute for oxygen production and carbon dioxide sequestration, this organisms are mainly composed and denominated as plankton [5].

Plankton comprises single-celled algae – phytoplankton (which realizes photosynthesis) - and generally small animals (mm or less) – zooplankton (secondary producers, herbivores), which are drifting in marine currents. Phytoplankton is responsible for about 45% of the global annual primary production and serve as food for zooplankton, which in its turn is an ideal size food for several commercially important fish and large aquatic mammals. Plankton is a vital component of marine and freshwater ecosystems. Besides, they also make important contributions to the global biogeochemical cycle and improve the accumulation of carbon dioxide in the atmosphere, 'pumping' carbon into the deepest regions of the sea [5].

Planktonic communities are frequently used as bioindicators to monitor ecological changes in aquatic ecosystems [6]. Thus, being a management tool to supervise the ecological system quality and to be a tool to take actions, for example to prevent algal blooms, toxic contamination from undisclosed source. This happens, because plankton reacts at the lowest variation of surrounding ecosystem. Plankton species and planktonic communities varies incited by many abiotic factors (light availability, temperature, salinity, heavy metals, pollutants, pH and nutrients concentration) and biotic factors (predators, parasites) [7]. These variations are being studied through ecological data to help policy makers, for example, where the plankton community varies and there is harmful plankton species that grows rapidly due the excessive nutrients in water [8].

However, the plankton interest is not only as ecological tool, but also holds industrial and biotechnological potential to be used in commercial products. Through an industrial perspective, phytoplankton and zooplankton species have been used as a feedstock for a wide range of applications, such as wastewater treatment, or production of high value compounds; and commercial products, such as food and feed supplements, pharmacological compounds, lipids, enzymes, biomass, polymers, toxins, pigments. Zooplankton is commonly used as live food for larval stages to the period of termination of fish, shrimp, mollusks and corals [9–11]. However, to exploit these organisms at a commercial and industrial level, there is a need to understand the ecological data to cultivate this organisms in a controlled methods to have a best effective method with reduced cost, due the impossible control in the wild ecosystems (where commercial exploitation provokes a negative impact) [12, 13].

This book chapter aims to analyze the several advantages that plankton, specifically phytoplankton and zooplankton, their qualities, ecological and economic relevance, as well as their cultivation techniques, aiming the production of add-value products.

2. Plankton ecological relevance

Plankton is a key-element to form the base of the aquatic food chain [14]. Every organism in the ocean habitat depends on plankton for their survival. Without them, the food chain will broke extensively provoking a shortage of the food basis [14]. For instance, bacterioplankton holds a key role to recycle compounds, minerals and energy within the food chain [15]. Due to climatic changes, plankton communities can change rapidly provoking diverse problems in the food chain, causing a bottom-up effect up to the fish, which is explored as a food source by humans. So, there is a need to monitor wild plankton communities to identify structural changes and, if necessary, to take actions in order to mitigate some of the negative changes, for example toxic algal blooms in marine ecosystems [4].

Plankton species are mostly short live forms and consequently, plankton communities are not greatly influenced by the persistence of older individuals from previous years. This can allow the joint of environmental changes and plankton dynamics, enabling fast analyzes unlike other aquatic organisms, such as fish species. Moreover, plankton can demonstrate dramatic changes within abiotic and biotic parameters variation (such as temperature, pH, salinity, nutrients and metals concentration, or even biotic changes, as bacteria or fungi proliferation) [16]. Regarding monitoring plankton communities, there are Continuous Plankton Recorders around the globe, aiming the development of studies about plankton dynamics (with abiotic and biotic data to understand plankton responses), and to contribute with updated data that will be pivotal to assist the management decisions of the stakeholders. In a large scale, this method has revealed itself, cost effective and essential to obtain data to understand the aquatic ecosystems [14].

2.1 Phytoplankton

Phytoplankton is one of the primary producers of the aquatic ecosystem, as well as the first organisms to produce energy, which they generate from light sources, such as solar. Phytoplankton converts light energy into carbohydrates through photosynthesis. The energy not auto consumed by them for survival and maintenance is available as food for herbivores or omnivores that feed on these microorganisms. Phytoplankton can absorb about 3% of the light energy that penetrate in the ocean. In fact, a low percentage when compared with terrestrial plants, which can absorb about 15% of the accessible sunlight. This divergence is triggered by the ocean itself, which absorbs sunlight in fluctuating grades. The sunlight is a limiting factor and a key source for phytoplankton survival and reproduction. If there is not enough sunlight, phytoplankton will diminish up to stable population [15].

2.2 Zooplankton

Zooplankton is composed by heterotrophic organisms that feed on phytoplankton, being mainly secondary consumers and aquatic herbivores. Thus, their energy is acquired from consuming the primary producers. The energy disposal is identical for tertiary consumers, as well as for phytoplankton, only the energy stored is available for predators. This predator can be a different zooplanktonic organism or a larger animal that grazes on plankton [15].

3. Specificities of the plankton

To fully understand plankton biotechnological potential, there is a need to evaluate their ecological specifications, according to the species and geographical habitat. Phytoplankton can be an useful and promising feedstock, due to their resilience and quick adaptation to environmental changes, which incontestably has consequences on their secondary metabolism [17].

3.1 Phytoplankton

There are evidences of the existence of microalgae since the Precambrian period, approximately 3.5 billion years ago. These microorganisms, mainly marine species, are responsible for the production and maintenance of atmospheric oxygen [18]. Algae have a fundamental role on ecological balance maintenance. Moreover they have a pivotal economic and social importance by supporting fauna, which is a source of food for humans [19] and other organisms [20].

Algae are considered a pool of several compounds with biological activities [21, 22]. The algal composition varies according to environmental conditions, thus there are species with different concentrations of proteins, polysaccharides, pigments and fatty acids [23].

Microalgae retains about 50% of carbon in their biomass, which is obtained in most cases from atmospheric carbon dioxide. Therefore, they are attracting interest for carbon sequestration in industrial processes [24, 25]. Nitrogen and phosphate compounds are essential nutrients for microalgae to protein and cell membrane synthesis. In this context, the application of microalgae in water bioremediation is a sustainable application to remove high amounts of these compounds from water bodies, mitigating their negative impacts [26].

3.2 Zooplankton

Zooplankton is offered as live food since the larval stages until the period of completion of fish, shrimp, mollusks and corals. They are organisms that have characteristics such as a rich nutritional composition, digestibility, buoyancy, ease of ingestion and attractive movement for post-larvae [27]. Rotifers are among the most widely used, mainly the genus *Brachionus* (Animalia, Monogononta), as an important source for the first zooplanktonic feeding for larvae of aquatic organisms, because they contemplate all the characteristics mentioned above, they have a high dietary value, being rich on polyunsaturated fatty acids and essentials amino acids, in addition to the appropriate size for the animal's feeding apparatus [28, 29].

Artemia or brine shrimp is an aquatic crustacean genus with nonselective feeding habit, which can feed on tiny particles of food like microalgae, bacteria, detritus and small organisms [30]. Artemia is a good model organism for ecotoxicological studies because they have a short life cycle and can be cultured in a large scale [31, 32].

The rotifers *Brachionus plicatilis* and *Brachionus rotundiformis* can be also cultivated at a large scale, meeting the demand for fish and shrimp larviculture [33]. Although they are considered a resource with a high nutritional value, it is important to note that this occurs due to the improvement of secondary cultivation techniques such as bioencapsulation, a technique in which the rotifer is enriched with foods with a high content of essential compounds, being fed for a time period less than 24 h and immediately offered in the larvae diet. Bioencapsulation allows rotifers to incorporate the nutritional characteristics of algae, subsequently transporting these elements to the fed larvae [34].

Copepods, used as live food, contribute to a better performance of fish larvae when compared to larvae fed with rotifers and Artemia [35, 36]. In general, copepod feeding results in an increase in survival, growth and a decrease in larval deformities [37, 38].

Due to a relatively high protein and nutrient content, *Moina* spp. (Branchiopoda, Cladocera) is a superior live food compared to *Artemia* [39, 40]. Cladocerans of the genus *Moina*, and *Moina macrocopa* in particular, are progressively important in aquaculture and ecotoxicology [41].

4. Plankton wild exploitation

There are commercial exploitation of plankton wild resources to provide marine food sources for human consumption, mainly zooplankton (example copepods and krill) [42]. This plankton presents a great economic potential because they are enriched biochemical profile, such lipids, proteins, pigments and other bioactive compounds. However, even at the lower food chain level they can accumulate heavy metals, organo-chlorides, dioxins and other harmful compounds, thus can be a problem if not analyzed rigorously [43]. However, at low quantities their risk is minimum when compared to higher food chain levels [44].

In this case, there are plankton specialized fisheries, where the harvest of the targeted species uses scientific data to harvest the adults in one specific season, with equipment to collect the plankton desired. For example, this happens in the Norwegian region from 1950 until today [44].

Although the plankton wild harvest needs a strong marine strategy to not cause environmental problems and to promote a sustainable plankton fishery, with reduced by-catch [44]. The economic importance and valorization are identical to the cultivated plankton, see Section 6. In this case, the most advantage is for animal feed due to: i- Greater diversity of organisms and possibility of compatibility with the larvae's and organism digestive apparatus; ii- The captured organisms will find themselves in different stages of development, and therefore, there must be some that have an adequate size to the requirements of apprehension of the cultivated larvae/organism; iii- The cost of capture is much lower than the cost of production of organisms used as live food. However, when compared to the cultivated, wild harvest demonstrates the consequent problems: i-the instable productivity rate due to the environment changes; ii- seasonality; iii-presence of parasite species, such as *Argulus* sp. e as *Lerneae* sp.; iv- maintenance of biochemical profile between harvests; v- possibility of accumulation of heavy metals, toxins, pollutants and harmful compounds.

5. Plankton cultivation

To avoid natural resources overexploitation, emerged the need to evolve plankton cultivation techniques. In this way, it is possible to produce enough biomass to supply industrial applications without putting pressure under marine ecosystems [45].

5.1 Phytoplankton

In aquaculture, microalgae serve as food and help to maintain water quality, as they produce oxygen, consume carbon dioxide and nitrogen compounds, especially ammonia [46]. In addition, they can still be used as bioindicators of the level of eutrophication of water bodies [47].

Microalgae are highly efficient photosynthetic organisms, and due to their high biotechnological potential, makes them one of the hot research topics of the moment [48]. Microalgal biomass can be commercially explored in different areas such as nutrition, human and animal, wastewater treatment, biodiesel production and to obtain compounds of interest to food, chemical and pharmaceutical industry [49, 50].

The main physico-chemical factors that affect the growth of microalgae are light, temperature, salinity and availability of nutrients [50].

Microalgae energy reserve substances consists in compounds of high molecular weight such as α-1,4 glucans, β-1,3 glucans and others of low molecular weight such as (glycosides and poly oils). In algae, the lipid reserve is needed for thee synthesis of lipoprotein membranes [51], and is also used to regulate the fluctuation of cells in water.

Lately, microalgae have been attracting the attention of researchers worldwide due to their resilience and high commercial interest [52].

The production of microalgal biomass, through photosynthetic growth, requires carbon dioxide, water, inorganic salts and temperatures generally between 20 to 30°C. To reduce the costs of microalgae biomass production, sunlight should be used, through outdoor cultivations, considering that the contamination is minimal, using essential nutrients such as nitrogen, phosphorus, iron and, in some cases, silica [49].

Currently, raceway ponds are the most used technique in the upscale production of microalgae to obtain biofuel. However, for this production to be more effective, technological advances must occur to develop photobioreactors which use light more efficiently, reducing the costs associated [53].

Microalgae cultivation is advantageous because it is possible to obtain metabolic products, which are used in feed of marine and terrestrial organisms, food supplements for humans, or for use in environmental processes, such as wastewater treatment, fertilization soil, biofuels and phytoremediation of toxic waste [54].

Species bioprospecting is very important to select the best strains that can produce the most desirable metabolic products. Several studies have evaluated the use of different microalgae for different purposes [55–57], but this field of research needs is currently evolving and much research still needs to be done.

Lourenço [58] reports that the interaction of microalgae with the culture medium and its physical environment results in significant changes in cell density, which tends to increase numerically in large proportions after inoculation. On the other hand, the concentrations of nutrients dissolved in the culture medium tend to decrease with their multiplication, reaching the point of complete exhaustion, depending on the time of development of the culture, stressing it.

The choice of the culture medium is extremely important for mass production of microalgae. Its improper use can affect the growth rate and the biochemical composition of cells [59, 60]. For each microalgae species, the productivity and the biochemical composition of the cells strongly depend on the type of cultivation and the nutrient profile of the medium [61].

According to Lourenço [58], the choice of the culture medium should consider the operational costs involved, since often low-cost culture media may be deficient in some components and do not allow the maximum production of algal biomass.

The microalgae possess various antioxidant properties and they are potential oxidative stress control alternatives in Artemia and, perhaps, other aquatic organisms used in aquaculture [62].

5.2 Zooplankton

Fiore and Tlusty [63] studied the incorporation of *Artemia* in commercial diets for larval diets of the American lobster (*Homarus americanus*) and found greater survival in stage IV post-larvae (19–25%) and subsequent juvenile performance when compared with a combination of *Artemia nauplii* with frozen *Artemia* incorporated in the diet. A diet 100% formulated resulted in reduced larval survival (6%) and post-larval size, while a larval diet of 100% of frozen adult *Artemia* resulted in reduced post-larval quality and early juvenile performance.

Vinh et al. [64] cite that the profitability of Artemia producing farms in the Mekong Delta, Vietnam, was significantly influenced by the geographic location and their interaction with the scale of production. To improve farm productivity, besides maintaining optimal stocking densities, moderate increases of organic fertilizer, feed and chemical inputs are recommended to supply Artemia with more nutrients and create better water environment for the optimal development and reproduction. Additionally, a periodic harvest of *Artemia* biomass (adult *Artemia*) is required to minimize food and space competition and provide more incomes to farmers.

Prusińska et al. [65] proved that the use of Artemia enriched in polyunsaturated fatty acids (PUFAs) in the larval cultivation of the freshwater fish (*Barbus barbus*), is an effective method to improve growth rates and feed utilization. Besides that, histological analyzes revealed better development of the active area of intestines, as well as an increase in the neutrophil count in the blood.

When cultivated, rotifers are relatively poor in eicosapentaenoic acid (EPA: 20: 5ω-3) and docosahexaenoic acid (DHA: 22: 6ω-3), and it is essential and therefore a common practice to enrich the culture with marine oil emulsions. Novel production techniques, such as closed recirculation systems are offering new possibilities for continuous supply of high-quality rotifers at densities 10 times greater than batch cultures. The increase in production in these systems is explained by the better water quality [66].

Yoshimura et al. [67] obtained a high density of rotifers (1.6×10^5 individuals mL^{-1}) using continuous filtration of water developed for ultra-high density production, equipped with a membrane filtration unit (pore size: 0.4 μm) and set inside a culture vessel. The culture performance of this system was tested by feeding with freshwater *Chlorella* (Chlorophyta) paste in a 4-day batch culture.

Alver et al. [68] used a system for automatic control of the growth and density of rotifer. The system computes feeding rates based on a setpoint for rotifer density and provides a fast growth period followed by rapid stabilization of the rotifer density. At the same time, overfeeding is prevented, thereby reducing the risk of cultivation crashes. Feeding rates are automatically computed based on measurements of the cultivation density and egg rate, and internal setpoints for growth rate and egg rate. The authors obtained densities in all tanks increasing from 60 to 90 mL^{-1} to the setpoint densities of 500 and 1000 mL^{-1} in 5–7 days, after insignificant growth on the first day. Gross growth rates slowed down considerably towards the end of the experiment, as the controller reduced feed rations in order to stabilize densities.

Han and Lee [69] studied the effects of salinity changes on the marine monogonont rotifer *Brachionus plicatilis* and found that a significant decrease in population growth was observed when the rotifers were grown in high salinity (35‰), leading to growth retardation and modulation of the antioxidant defense system. These findings provide a better understanding on the adverse effects of salinity changes on lifecycle parameters and oxidative stress defense mechanism in rotifers.

Chilmawati and Suminto [70] observed the performance of copepod *Oithona* sp. in different diets with microalgae *Chaetoceros calcitrans* (Bacillariophyta), *Chlorella vulgaris*, *Nannochloropsis oculata* (Ochrophyta, Eustigmatophyceae) and *Isochrysis galbana* (Haptophyta, Coccolithophyceae). The results showed that the diet of phytoplankton cells was significantly different in the growth performance of *Oithona* sp. The diet of *C. calcitrans* gave the best growth performance of *Oithona* sp., when reached 6,963 ± 0.38 ind mL^{-1} of total density (0.121 ± 0.003) and specific growth rate and egg production (16.50 ± 2.74 ind^{-1}).

Knuckey et al. [71] cultivated the copepod *Acartia sinjiensis* in a variety of mono and binary algal diets and observed that there were significant differences in the rate of development of copepods between diets. *Rhodomonas* (Cryptophyta) was confirmed as an excellent algal diet for *Acartia* (Crustacea, Copepoda), but it is often unpredictable in mass culture. The cryptophyte, Cryptomonad sp. (CS-412) showed to support an equally rapid development rate with the advantage of being more stable in mass culture. The algal feed concentration for maximal copepod development rate was dependent on the algal feed species.

Puello-Cruz et al. [72] cultivated the copepod *Pseudodiaptomus euryhalinus* (Crustacea, Copepoda) in a mono-microalgae culture (*Chaetoceros muelleri*, *Nannochloropsis oculata*, *Isochrysis galbana*, *Tetraselmis suecica* (Chlorophyta), or a commercial frozen concentrate of *Tetraselmis* sp.) and in binary diets (*C. muelleri*: *I. galbana* in 1: 1 and 2: 1 cell ratios and *C. muelleri*: *I. galbana*: frozen *Tetraselmis* sp. in 2: 2: 1 ratio). These gave better results than the cultures of *N. oculata*, *I. galbana*, *T. suecica* and the frozen *Tetraselmis* concentrate, but the production was similar to that obtained with *C. muelleri* supplied as a monoalgal diet, showing that the addition of *C. muelleri* may improve the performance of other monoalgal diets, whereas the addition of other microalgae is unlikely to improve the results obtained when *C. muelleri* is supplied as a monoalgal diet.

Using relatively simple culture techniques, in transparent plastic boxes (32 × 47 × 14.5 cm) containing 4.5 L of filtered aerated seawater at room temperature (28 to 32°C) and a salinity of 35‰, Ribeiro and Souza-Santos [73] cultivated the copepod *Tisbe biminiensis* fed with commercially available ornamental fish food and every two days following water exchange, with 500 mL of one of the following diatoms: *Phaeodactylum tricornutum* or *Thalassiosira fluviatilis* (Bacillariophyta). The collection of *T. biminiensis* from the 5 L cultures produced a mean of 28,000 nauplii and copepodites L^{-1} day^{-1} over a 130-day period.

Sarkisian et al. [74] used an innovative design for an intensive culture system of the calanoid copepod *Acartia tonsa*, a prime candidate for use as a live food item. The system output was on average 22 million eggs day^{-1} (21,955,420 ± 8,709,668) with an average hatch rate of 49% (49.1 ± 14.8) over three seasons.

Poynton et al. [41] cultivated females of the cladoceran *Moina macrocopa* in a situation of flagellate infection associated with mortality. At day 10, all *M. macrocopa* were alive in uninfected cultures, whereas in untreated infected cultures, the survival was significantly lower: only 26% of cladocerans were alive. In infected cultures treated with humic substances (25 mg L^{-1} of dissolved organic carbon), mortalities were comparable to those in the untreated infected cultures; in contrast, in the infected cultures treated with 4 g L^{-1} sea salt, mortalities were interrupted, and 76% of the *M. macrocopa* were alive at day 10.

Liu et al. [75] studied the effects of a polystyrene nanoplastic on physiological changes (e.g., survival, growth, and reproduction) and expression levels of stress defense genes (oxidative stress-mediated and heat shock proteins) in the freshwater flea *Daphnia pulex*. The results showed that the digestive organs of *D. pulex* were strongly fluorescent after exposure to the nanoplastic particles and the 48 h median lethal concentration (LC_{50}) of the nanoplastic was determined to be 76.69 mg L^{-1}. The time to brood was delayed, and total offspring per female and number of broods were decreased in all the treatment groups. In addition, the offspring per brood were significantly decreased in the 0.1 mg L^{-1} group.

Raymundo et al. [76] compared the sensitivity of temperate and tropical cladocerans to different insecticides. The order of sensitivity of the native cladocerans to chlorpyrifos was: *Ceriodaphnia silvestrii* (0.039 μg L^{-1}) > *Diaphanosoma birgei* (0.211 μg L^{-1}) = *Daphnia laevis* (0.216 μg L^{-1}) > *Moina micrura* (0.463 μg L^{-1}) = *Macrothrix flabelligera* (0.619 μg L^{-1}). A regulatory acceptable concentration based on temperate cladoceran toxicity data of both chlorpyrifos and other insecticides also appeared to be sufficiently protective for tropical cladoceran species.

Jaikumar et al. [77] described that the sensitivity to microplastics can differ between different species of cladocerans and can be drastically influenced by the temperature, although in high concentrations of exposure.

Hansen [78] cultivated the planktotrophic larvae of the boreal capitellid polychaete *Mediomastus fragile*, fed with the microalgae *Isochrysis galbana* and concluded that the larvae were able to capture and ingest particles in the size spectrum between 2 and 10 μm. However, the optimal particle size was 7 μm. The larvae enter the plankton in the early spring, when the phytoplankton size spectrum is typically dominated by large algal cells, exceeding the size for efficient uptake. The physical limitations for particle capture are therefore a potential limit for feeding. The ability to delay larval development is an advantage for a planktotrophic larvae functioning as a growing dispersive organism.

6. Plankton economic importance

In diverse industry areas, microalgae have been widely used as a source for a variety of practices and potential metabolic products, such as food supplements, pharmacological substances, lipids, enzymes, biomass, polymers, toxins, pigments or tertiary sewage treatment. They are also important in aquaculture as a source of nutrients and are of great importance in the production of oxygen, carbon dioxide sequestration and nitrogenous compounds removal, such as ammonia [46, 54, 58]. They are also used as bioindicators, reporting water bodies ecological quality status [47]. However, it is considered that the plankton biotechnology is still young when compared to macroalgal and terrestrial plant biotechnological exploitation and knowledge [79]. Nevertheless, when compared with this two biotechnology branches, it is estimated that plankton have specimens and more suitable, due to their reduced form, being mainly aquatic, a life cycle shortened and rapid adaptation of the metabolism which is capable to produce various interesting compounds [9, 13, 80].

The production of microalgae in different sectors generates social, environmental and economic benefits. For example, in the USA and India, *Haematococcus lacustris* (formerly *Haematococcus pluvialis*) (Chlorophyta) production aims the extraction of astaxanthin, used as a food coloring and also as a powerful antioxidant in the pharmaceutical and cosmetic industry [81].

According to Wijffels [53], marine biotechnology aims to discover new products that can contribute to the health of human beings, such as, for example, new nutraceuticals obtained from algae for use in human and animal feed industries, besides the contribution also in the energy sector, such as the production of biofuels. According to the author, the ω-3 fatty acids, provenly beneficial for human health, can also be a potential source of biofuels. Therefore, the biggest challenge is to obtain these products with quality, in enough quantities and in a sustainable way.

6.1 Food and nutraceuticals

The search to food sources are advancing as an indispensable resolve the feed problem, with the continuous world's population grow restricted, by the global restrictions [82]. Phytoplankton aquaculture in an industrial large-scale to human food usage begin with the cultivation of *Chlorella vulgaris* during World War II [80].

According to Pulz and Gross [79], the functional food market using microalgae, in pasta, breads, yoghurts and beverages, is rapidly developing in countries, such as France, United States, China and Thailand. The most common application has been in aquaculture, for the direct or indirect feeding of some species of fish, mollusks, crustaceans and other organisms of economic interest [83].

The consumption of ω-3 obtained from microalgae is beneficial for neural development, in addition to preventing coronary problems, cancer, hypertension, diabetes, cystic fibrosis, arthritis, asthma, schizophrenia and depression. Marine microalgae are capable of synthesizing ω-3 fatty acids, eicosapentaenoic (EPA, C20: 5) and docosahexaenoic (DHA, C22: 6), which enter the marine food chain and are available in fish oil. These fatty acids are considered important in the development of brain tissue and visual function [84].

Microalgae are the main producers of biomass that accumulate in higher organisms through the food chain. For several centuries, they are used as food in Southeast Asian countries, mainly due to their high protein content. Recently, microalgae have attracted the interest of many researchers due to their structurally diverse bioactive compounds, efficient photosynthetic machinery, greater mass productivity and the absence of competition with arable land and drinking water. They can withstand adverse environmental conditions, producing a variety of biologically active primary and secondary metabolites, such as polysaccharides, carotenoids, omega-3 and 6 fatty acids and phenolic compounds. These metabolites exhibit a series of pharmacological activities, which include therapeutic, drug-carrying and physiochemical properties, including gelation, swelling and emulsification. These may be a new source of functional compounds in the food and pharmaceutical industries [85].

Currently, microalgae are being incorporated into many food formulations. Most of them use microalgae as a marketing strategy or as a coloring agent. As for example, the cyanobacterium Spirulina is not only in fashion, but is rich in several valuable and highly nutritious compounds, such as proteins, PUFAs and bioactive pigments, including chlorophylls, carotenoids and phycobiliproteins. One of the main advantages of natural pigments derived from Spirulina, when compared to their synthetic counterparts, is that the former has several health benefits, and can be used as an ingredient in the development of new functional foods. Proteins from Spirulina have proven to be excellent sources of bioactive peptides with potential application in the functional food industry as antihypertensive, anti-diabetic, anti-obesity and antioxidant ingredients [86] immunomodulatory and anti-inflammatory among other positive bioactivities [87].

Some of the prerequisites for using algae biomass for humans and animals include determining the chemical composition; toxic biogenic substances; non-biogenic toxic compounds; protein quality studies; biochemical nutritional studies; supplemental value of algae to conventional food sources; health analysis; safety assessments (animal feeding tests); clinical studies (safety test and suitability of the product for human consumption) and acceptability studies [88].

The microalgae used as a food supplement are generally sold in the form of tablets, capsules and liquids or are incorporated in pasta, snacks, candy bars, ice cream, chewing gum, in mixtures of drinks and dyes for natural foods [88, 89]. Foods supplemented with microalgae biomass, when properly processed, can make foods more colorful and tasty, adding not only nutritional value, but also new, unique and attractive flavors [50].

The reasons for this recent growth in interest are cost-effective cultivation and a short cultivation time until the desired compost is obtained. In addition, they have the status generally considered safe and as such do not contain any toxins or pathogens that can be transmitted to humans. [90].

6.2 Feed

The plankton is a natural source for various animals' species, which are cultivated. Consequently, they are a standard feed source to various farmed species. To other animals, they are non-natural feed source, which is used supplement to be incorporated with normal feed, similarly the plankton usage as human food supply, due to the high quality of protein, minerals, vitamins, carbohydrates and also essential fatty acids to be a high quality feed for fish and others animals [91].

Phytoplankton is a vital player in aquaculture (mariculture) as they are the natural food bases to larvae life stage of various types of mollusks, crustaceans, and fish. The utmost phytoplankton used in aquaculture worldwide belong to the genera: *Chlorella*, *Tetraselmis*, *Isochrysis*, *Pavlova* (Haptophyta, Pavlovophyceae), *Phaeodactylum*, *Chaetoceros*, *Nannochloropsis*, *Skeletonema*, and *Thalassiosira* (Bacillariophyta) [92].

The use of plankton as feed improver was attainment further attention by the I&D research teams and industry to develop feeds to diverse animals (mainly in aquaculture). Which, the main results are the animals feed with plankton gain weight, enhance of triglyceride profile and the protein deposition in muscle, the animal digestibility, starvation tolerance and carcass quality [91, 93].

Phytoplankton can be cast-off as a source of natural pigments for the culture of prawns, salmonid fish, and ornamental fish [91].

6.3 Cosmetic

The cosmetic area is the third major commercial segment for phytoplankton application, due to the research of natural products to substitute synthetic ingredients. Thus, with cosmetic consumers turning their mindset, the cosmetic segment is one of the main actives to explore the biotechnological potential of the plankton. The natural and ecofriendly predispositions in this area, give an new input to find new high value, innovative and natural formulations for new products, without the imposition of reduced costs as the other areas [80]. The microalgae were not very common in cosmetic, nonetheless, microalgae and their derivatives are in beginning to be integrated in diverse formulas to skin and hair products, through a wide range of functions, such as excipient (stabilizer or emulsifier) or active ingredient. The phytoplankton is usually used in moisturizing, skin whitening, anti-aging, and sun protection creams formulations. However, the pigments from phytoplankton is cast-off as colorant agent for varied cosmetic products [94].

6.4 Bioremediation

The application of microalgae to bioremediate wastewaters shows a great potential to complement traditional wastewater treatment processes. Furthermore, this approach addresses the need to reduce the costs associated with the growth media expenses for microalgae biomass production [95], through wastewater recycling to obtain microalgal biomass instead of culture medium [96].

Nevertheless, it is necessary to consider possible sources of growth medium contamination, such as grazers which feed on microalgae (**Figure 1a** and **b**), as well as the presence of other microalgae species that can compete or inhibit the target species production.

Bioremediation of numerous pollutants of different characteristics and properties released from the domestic, industrial, agricultural and aquaculture sectors [97, 98]. Moreover, promoting microalgae cultivation in wastewater will help mitigate the environmental impacts of treated effluents since this biological method will complement conventional wastewater treatment and improve not only the removal of organic and inorganic load but also the removal of emerging pollutants, such as pesticides, metals, pharmaceuticals or household cleaning chemicals [99–102].

In addition, they are also capable of removing metals, incorporating them in their cell wall [103] and other noxious compounds such as phenols and chlorophenols [104].

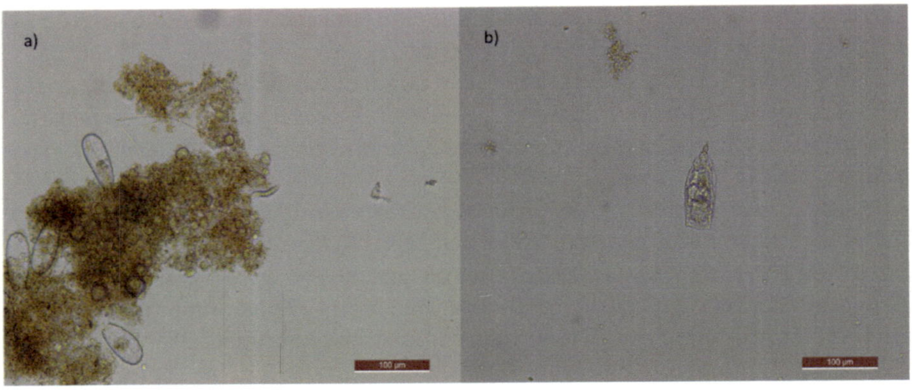

Figure 1.
Microscopic observations of Chlorella vulgaris cultivation in municipal wastewater sludge centrate, (a) with the presence of other microalgae species and (b) with the presence of grazers.

6.5 Renewables energies

An emerging area for microalgae biotechnology is environmental applications. This is mainly due to its carbon dioxide mitigation capacity, reducing greenhouse gas emissions that are related to global warming and climate change; and its ability to grow in an effluent liquid that allows wastewater treatment. Today, there is a focus on the use of microalgae in renewable energy as a potential source for the production of biofuels, such as biodiesel, bioethanol, biohydrogen and biogas [105].

It is worth mentioning the importance of the production of biofuels through microalgae. Microalgae naturally contain about 10% lipids. These lipids are mainly present in photosynthetic membranes. Microalgae accumulate lipids in high concentration under "stress" conditions, caused, for example, by the depletion of nutrients such as nitrogen. In the absence of these nutrients, growth is hampered, while energy is continuously received in the form of light. Microalgae channel excess energy into large macromolecules, such as lipids or starch. In these cases, the lipid content can reach 60%. Under stressful conditions, these lipids accumulate in body lipids such as triacylglycerides or neutral lipids. The neutral lipids can be used as raw material for the production of biofuels [106].

During the past few decades, many research studies have covered different technologies to produce biodiesel from lipid-rich microalgae. Under controlled cultivation conditions, microalgae can accumulate metabolites intended to produce various biofuels. For example, starch and various types of oils can be bioaccumulated. Starch extracted from algae is easily hydrolyzed to glucose and used for fermentation in the production of bioethanol. Currently, commercial production of bioethanol from algae is not a viable choice due to the low yield of the product compared to other terrestrial biomasses. The high costs of algae cultivation systems are due to several complex steps: (i) algae cultivation; (ii) harvest; (iii) pre-treatment of biomass; (iv) fermentation; and (v) extraction of bioethanol. By linking all possible improvements at each stage of the process, a substantial advance towards cost-effective algae systems can be achieved in the future [107].

7. Conclusions

This chapter covered the many advantages that plankton have, specifically phytoplankton and zooplankton, their qualities, ecological and economic relevance, as well as their cultivation techniques, aiming the production of add-value products with industrial interest.

It is of great need to use all the knowledge presented and apply it in the different branches of ecology, industry or science, aiming the discovery of new products or directing it to a specific study area, being a subsidy of great importance for the environment and/or for the human being.

Acknowledgements

This work is financed by national funds through FCT - Foundation for Science and Technology, I.P., within the scope of the projects UIDB/04292/2020 – MARE - Marine and Environmental Sciences. This work was financed by the Live Food Production Laboratory (LABPAV) and the Tropical Aquaculture Study Group (GEAQUI) of the Federal Institute of Education, Science and Technology of Ceará —IFCE, Campus Aracati, Ceará, Brazil. João Cotas thanks to the European Regional Development Fund through the Interreg Atlantic Area Program, under the project NASPA (EAPA_451/2016). Diana Pacheco thanks the PTDC/BIA-CBI/31144/2017—POCI-01project -0145-FEDER-031144—MARINE INVADERS, co-financed by the ERDF through POCI (Operational Program Competitiveness and Internationalization) and by the Foundation for Science and Technology (FCT, IP).

Conflict of interest

The authors declare no conflict of interest.

Author details

Glacio Souza Araujo[1], Diana Pacheco[2], João Cotas[2], José William Alves da Silva[1], Jefferson Saboya[3], Renato Teixeira Moreira[1] and Leonel Pereira[2*]

1 Ceará Federal Institute/IFCE-Campus Aracati, Aracati, Brazil

2 Department of Life Sciences, MARE-Marine and Environmental Sciences Centre, University of Coimbra, Coimbra, Portugal

3 EEEP Marta Maria Giffoni de Sousa, Acaraú, Brazil

*Address all correspondence to: leonel.pereira@uc.pt

References

[1] Miller CB, Wheeler PA. Biological Oceanography, 2nd Edition | Wiley. Hoboken, NJ, USA: Wiley-Blackwell; 2012.

[2] Field JG, Hempel G, Summerhayes CP, editors. Oceans 2020: Science, Trends, and the Challenge of Sustainability Illustrated Edition. Washington, D.C., USA: Island Press; 2002.

[3] Costanza R, D'Arge R, de Groot R, Farber S, Grasso M, Hannon B, et al. The value of the world's ecosystem services and natural capital. Nature 1997;387:253-260. https://doi.org/10.1038/387253a0.

[4] Hays G, Richardson A, Robinson C. Climate change and marine plankton. Trends Ecol Evol 2005;20:337-344. https://doi.org/10.1016/j.tree.2005.03.004.

[5] Brierley AS. Plankton. Curr Biol 2017;27:R478–R483. https://doi.org/10.1016/j.cub.2017.02.045.

[6] Paul S, Wooldridge T, Perissinotto R. Evaluation of abiotic stresses of temperate estuaries by using resident zooplankton: A community vs. population approach. Estuar Coast Shelf Sci 2016;170:102-111. https://doi.org/10.1016/j.ecss.2016.01.007.

[7] Wiltshire KH, Boersma M, Carstens K, Kraberg AC, Peters S, Scharfe M. Control of phytoplankton in a shelf sea: Determination of the main drivers based on the Helgoland Roads Time Series. J Sea Res 2015;105:42-52. https://doi.org/10.1016/j.seares.2015.06.022.

[8] Anderson DM, Glibert PM, Burkholder JM. Harmful algal blooms and eutrophication: Nutrient sources, composition, and consequences. Estuaries 2002;25:704-726. https://doi.org/10.1007/BF02804901.

[9] Olaizola M, Grewe C. Commercial Microalgal Cultivation Systems. In: Hallmann A, Rampelotto P, editors. Gd. Challenges Algae Biotechnol., Cham, Switzerland: Springer; 2019, p. 3-34. https://doi.org/10.1007/978-3-030-25233-5_1.

[10] Vigani M, Parisi C, Rodríguez-Cerezo E, Barbosa MJ, Sijtsma L, Ploeg M, et al. Food and feed products from micro-algae: Market opportunities and challenges for the EU. Trends Food Sci Technol 2015;42:81-92. https://doi.org/10.1016/j.tifs.2014.12.004.

[11] Voort MPJ van der, Vulsteke E, Visser CLM de. Macro-economics of Algae products, Public Output report WP2A7.02 of the EnAlgae project. Swansea, UK: 2015.

[12] Rahman KM. Food and High Value Products from Microalgae: Market Opportunities and Challenges. In: Alam MA, Xu J-L, Wang Z, editors. Microalgae Biotechnol. Food, Heal. High Value Prod., Singapore: Springer Singapore; 2020, p. 3-27. https://doi.org/10.1007/978-981-15-0169-2_1.

[13] Enzing C, Ploeg M, Barbosa M, Sijtsma L. Microalgae-based products for the food and feed sector: an outlook for Europe. 2014. https://doi.org/10.2791/3339.

[14] Batten SD, Abu-Alhaija R, Chiba S, Edwards M, Graham G, Jyothibabu R, et al. A Global Plankton Diversity Monitoring Program. Front Mar Sci 2019;6. https://doi.org/10.3389/fmars.2019.00321.

[15] NMFS-COPEPOD: Why are plankton important? n.d. https://www.st.nmfs.noaa.gov/plankton/intro/why.html (accessed November 12, 2020).

[16] Taylor AH, Allen JI, Clark PA. Extraction of a weak climatic signal by

an ecosystem. Nature 2002;416:629-632. https://doi.org/10.1038/416629a.

[17] Sonnenschein EC, Pu Y, Beld J, Burkart MD. Phosphopantetheinylation in the green microalgae *Chlamydomonas reinhardtii*. J Appl Phycol 2016;28:3259-3267. https://doi.org/10.1007/s10811-016-0875-7.

[18] Tahmasebi A, Kassim MA, Yu J, Bhattacharya S. Thermogravimetric study of the combustion of *Tetraselmis suecica* microalgae and its blend with a Victorian brown coal in O2/N2 and O2/CO2 atmospheres. Bioresour Technol 2013;150:15-27. https://doi.org/10.1016/j.biortech.2013.09.113.

[19] Gaffney M, O'Rourke R, Murphy R. Manipulation of fatty acid and antioxidant profiles of the microalgae *Schizochytrium* sp. through flaxseed oil supplementation. Algal Res 2014;6:195-200. https://doi.org/10.1016/j.algal.2014.03.005.

[20] El-Sayed HS, Ibrahim HAH, Beltagy EA, Khairy HM. Effects of short term feeding of some marine microalgae on the microbial profile associated with *Dicentrarchus labrax* post larvae. Egypt J Aquat Res 2014;40:251-260. https://doi.org/10.1016/j.ejar.2014.08.001.

[21] da Silva JWA, Araújo GS, Coelho AA da C, Bezerra JHC, Lopes DNM, Farias WRL. Effect of nitrate depletion on lipid accumulation by the marine microalga *Nannochloropsis oculata*. Bol Do Inst Pesca 2015;41:811-816. https://doi.org/10.20950/1678-2305.2015v41nep811.

[22] Pacheco D, Rocha AC, Pereira L, Verdelhos T. Microalgae Water Bioremediation: Trends and Hot Topics. Appl Sci 2020;10:1886. https://doi.org/10.3390/app10051886.

[23] George B, Pancha I, Desai C, Chokshi K, Paliwal C, Ghosh T, et al. Effects of different media composition, light intensity and photoperiod on morphology and physiology of freshwater microalgae *Ankistrodesmus falcatus* – A potential strain for bio-fuel production. Bioresour Technol 2014;171:367-374. https://doi.org/10.1016/j.biortech.2014.08.086.

[24] Biller P, Ross AB. Pyrolysis GC–MS as a novel analysis technique to determine the biochemical composition of microalgae. Algal Res 2014;6:91-97. https://doi.org/10.1016/j.algal.2014.09.009.

[25] Raeesossadati MJ, Ahmadzadeh H, McHenry MP, Moheimani NR. CO_2 bioremediation by microalgae in photobioreactors: Impacts of biomass and CO_2 concentrations, light, and temperature. Algal Res 2014;6:78-85. https://doi.org/10.1016/j.algal.2014.09.007.

[26] Chang Z, Duan P, Xu Y. Catalytic hydropyrolysis of microalgae: Influence of operating variables on the formation and composition of bio-oil. Bioresour Technol 2015;184:349-354. https://doi.org/10.1016/j.biortech.2014.08.014.

[27] Páez-Osuna F. Retos y perspectivas de la camaronicultura en la zona costera. Rev Latinoam Recur Nat 2005;1:21-31.

[28] Planas M, Vázquez JA, Marqués J, Pérez-Lomba R, González MP, Murado M. Enhancement of rotifer (*Brachionus plicatilis*) growth by using terrestrial lactic acid bacteria. Aquaculture 2004;240:313-329. https://doi.org/10.1016/j.aquaculture.2004.07.016.

[29] Rojo-Cebreros AH, Ibarra-Castro L, Guerrero-Carlock E, Luis Sánchez-Téllez J, Alvarez-Lajonchère L. Pilot-scale production of the rotifer *Brachionus* sp. under different culture systems. Rev Biol Mar Oceanogr 2017;52:539-549. https://doi.org/10.4067/S0718-19572017000300011.

[30] Sorgeloos P, Lavens P, Leger P, Tackaert W, Versichele D. Manual for the culture and use of brine shrimp Artemia in aquaculture. Artemia Reference Center, State University of Ghent, Ghent, Belgium; 1986.

[31] Ikhwanuddi M, Azra MN, Sung YY, Munafi ABA-, Shabdin ML. Live Foods for Juveniles' Production of Blue Swimming Crab, *Portunus pelagicus* (Linnaeus, 1766). J Fish Aquat Sci 2012;7:266-278. https://doi.org/10.3923/jfas.2012.266.278.

[32] Sung YY, Pineda C, MacRae TH, Sorgeloos P, Bossier P. Exposure of gnotobiotic *Artemia franciscana* larvae to abiotic stress promotes heat shock protein 70 synthesis and enhances resistance to pathogenic *Vibrio campbellii*. Cell Stress Chaperones 2008;13:59-66. https://doi.org/10.1007/s12192-008-0011-y.

[33] Lubzens E, Zmora O. Production and Nutritional Value of Rotifers. In: Støttrup JG, McEvoy LA, editors. Live Feed. Mar. Aquac., Oxford, UK: Blackwell Science Ltd; 2007, p. 17-64. https://doi.org/10.1002/9780470995143.ch2.

[34] Ferreira PMP. Manual de cultivo e bioencapsulação da cadeia alimentar para a larvicultura de peixes marinhos. Lisbon, Portugal: Instituto Nacional de Recursos Biológicos IP - IPIMAR; 2009.

[35] Abate TG, Nielsen R, Nielsen M, Jepsen PM, Hansen BW. A cost-effectiveness analysis of live feeds in juvenile turbot *Scophthalmus maximus* (Linnaeus, 1758) farming: copepods versus Artemia. Aquac Nutr 2016;22:899-910. https://doi.org/10.1111/anu.12307.

[36] Øie G, Galloway T, Sørøy M, Holmvaag Hansen M, Norheim IA, Halseth CK, et al. Effect of cultivated copepods (*Acartia tonsa*) in first-feeding of Atlantic cod (*Gadus morhua*) and ballan wrasse (*Labrus bergylta*) larvae. Aquac Nutr 2017;23:3-17. https://doi.org/10.1111/anu.12352.

[37] Busch KET, Peruzzi S, Tonning F, Falk-Petersen I-B. Effect of prey type and size on the growth, survival and pigmentation of cod (*Gadus morhua* L.) larvae. Aquac Nutr 2011;17:e595–e603. https://doi.org/10.1111/j.1365-2095.2010.00800.x.

[38] Wilcox JA, Tracy PL, Marcus NH. Improving Live Feeds: Effect of a Mixed Diet of Copepod Nauplii (*Acartia tonsa*) and Rotifers on the Survival and Growth of First-Feeding Larvae of the Southern Flounder, *Paralichthys lethostigma*. J World Aquac Soc 2006;37:113-120. https://doi.org/10.1111/j.1749-7345.2006.00014.x.

[39] Alam MJ, Ang KJ, Cheah SH. Use of Moina micrura (Kurz) as an Artemia substitute in the production of *Macrobrachium rosenbergii* (de Man) post-larvae. Aquaculture 1993;109:337-349. https://doi.org/10.1016/0044-8486(93)90173-V.

[40] Loh JY, Ong HKA, Hii YS, Smith TJ, Lock MW, Khoo G. Highly Unsaturated Fatty Acid (HUFA) Retention in the Freshwater Cladoceran, *Moina macrocopa*, Enriched With Lipid Emulsions. Isr J Aquac - BAMIGDEH 2012;64:1-9.

[41] Poynton SL, Dachsel P, Lehmann MJ, Steinberg CEW. Culture of the cladoceran *Moina macrocopa*: Mortality associated with flagellate infection. Aquaculture 2013;416-417:374-379. https://doi.org/10.1016/j.aquaculture.2013.09.029.

[42] Skjoldal HR, editor. The Norwegian Sea Ecosystem. Trondheim, Norway: Tapir Academic Press; 2005.

[43] Mizukawa K, Takada H, Takeuchi I, Ikemoto T, Omori K, Tsuchiya K. Bioconcentration and biomagnification

of polybrominated diphenyl ethers (PBDEs) through lower-trophic-level coastal marine food web. Mar Pollut Bull 2009;58:1217-1224. https://doi.org/10.1016/j.marpolbul.2009.03.008.

[44] Grimaldo E, Helge S. Commercial Exploitation of Zooplankton in the Norwegian Sea. Funct. Ecosyst., InTech; 2012. https://doi.org/10.5772/36099.

[45] Deruyck B, Hue K, Nguyen T, Decaestecker E, Muylaert K. Modeling the impact of rotifer contamination on microalgal production in open pond, photobioreactor and thin layer cultivation systems. Algal Res 2019;38:101398. https://doi.org/10.1016/j.algal.2018.101398.

[46] Derner RB. Cultivo de microalgas. Produção camarão Mar., Florianópolis, Brazil: UFSC; 1996, p. 64-75.

[47] Trobajo R, Cox EJ, Quintana XD. The effects of some environmental variables on the morphology of *Nitzschia frustulum* (Bacillariophyta), in relation its use as a bioindicator. Nov Hedwigia 2004;79:433-445. https://doi.org/10.1127/0029-5035/2004/0079-0433.

[48] Miao X, Wu Q. Biodiesel production from heterotrophic microalgal oil. Bioresour Technol 2006;97:841-846. https://doi.org/10.1016/j.biortech.2005.04.008.

[49] Grobbelaar JU. Algal Nutrition - Mineral Nutrition. In: Richmond A, editor. Handb. Microalgal Cult., Oxford, UK: Blackwell Publishing Ltd; 2007, p. 95-115. https://doi.org/10.1002/9780470995280.ch6.

[50] Richmond A. Handbook of Microalgal Culture: Biotechnology and Applied Phycology | Wiley. Hoboken, NJ, USA: Blackwell Science Ltd; 2004.

[51] Lee RF, Valkirs AO, Seligman PF. Importance of microalgae in the biodegradation of tributyltin in estuarine waters. Environ Sci Technol 1989;23:1515-1518. https://doi.org/10.1021/es00070a010.

[52] Milledge JJ. Commercial application of microalgae other than as biofuels: a brief review. Rev Environ Sci Bio/Technology 2011;10:31-41. https://doi.org/10.1007/s11157-010-9214-7.

[53] Wijffels RH. Potential of sponges and microalgae for marine biotechnology. Trends Biotechnol 2008;26:26-31. https://doi.org/10.1016/j.tibtech.2007.10.002.

[54] Perez-Garcia O, Escalante FME, De-Bashan LE, Bashan Y. Heterotrophic cultures of microalgae: Metabolism and potential products. Water Res 2011;45:11-36. https://doi.org/10.1016/j.watres.2010.08.037.

[55] Francisco ÉC, Neves DB, Jacob-Lopes E, Franco TT. Microalgae as feedstock for biodiesel production: Carbon dioxide sequestration, lipid production and biofuel quality. J Chem Technol Biotechnol 2010;85:395-403. https://doi.org/10.1002/jctb.2338.

[56] Yoo C, Jun S-Y, Lee J-Y, Ahn C-Y, Oh H-M. Selection of microalgae for lipid production under high levels carbon dioxide. Bioresour Technol 2010;101:S71–S74. https://doi.org/10.1016/j.biortech.2009.03.030.

[57] Mutanda T, Ramesh D, Karthikeyan S, Kumari S, Anandraj A, Bux F. Bioprospecting for hyper-lipid producing microalgal strains for sustainable biofuel production. Bioresour Technol 2011;102:57-70. https://doi.org/10.1016/j.biortech.2010.06.077.

[58] Lourenço SO. Cultivo de microalgas marinhas: princípios e aplicações. São Carlos, Brazil: RiMa; 2006.

[59] Sánchez S, Martínez M, Espinola F. Biomass production and biochemical

variability of the marine microalga *Isochrysis galbana* in relation to culture medium. Biochem Eng J 2000;6:13-18. https://doi.org/10.1016/S1369-703X(00)00071-1.

[60] Lourenço E, Marques JAN. Produção primária marinha. In: Pereira RC, Soares-Gomes A, editors. Biol. Mar., Rio de Janeiro, Brazil: Interciência; 2002, p. 195-227.

[61] Guedes AC, Amaro HM, Malcata FX. Microalgae as sources of high added-value compounds-a brief review of recent work. Biotechnol Prog 2011;27:597-613. https://doi.org/10.1002/btpr.575.

[62] Tiong IKR, Nagappan T, Abdul Wahid ME, Tengku Muhammad TS, Tatsuki T, Satyantini WH, et al. Antioxidant capacity of five microalgae species and their effect on heat shock protein 70 expression in the brine shrimp Artemia. Aquac Reports 2020;18:100433. https://doi.org/10.1016/j.aqrep.2020.100433.

[63] Fiore DR, Tlusty MF. Use of commercial Artemia replacement diets in culturing larval American lobsters (*Homarus americanus*). Aquaculture 2005;243:291-303. https://doi.org/10.1016/j.aquaculture.2004.10.009.

[64] Vinh NP, Huang CT, Hsiao YJ, Hieu TK, Chen LH. Data envelopment analysis for production efficiency improvement: An empirical application on brine shrimp *Artemia franciscana* culture in the Mekong Delta, Vietnam. Aquac Res 2020;51:2985-2996. https://doi.org/10.1111/are.14636.

[65] Prusińska M, Nowosad J, Jarmołowicz S, Mikiewicz M, Duda A, Wiszniewski G, et al. Effect of feeding barbel larvae (*Barbus barbus* (L, 1758)) *Artemia* sp. nauplii enriched with PUFAs on their growth and survival rate, blood composition, alimentary tract histological structure and body chemical composition. Aquac Reports 2020;18:100492. https://doi.org/10.1016/j.aqrep.2020.100492.

[66] Dhert P, Rombaut G, Suantika G, Sorgeloos P. Advancement of rotifer culture and manipulation techniques in Europe. Aquaculture 2001;200:129-146. https://doi.org/10.1016/S0044-8486(01)00697-4.

[67] Yoshimura K, Tanaka K, Yoshimatsu T. A novel culture system for the ultra-high-density production of the rotifer, *Brachionus rotundiformis*—a preliminary report. Aquaculture 2003;227:165-172. https://doi.org/10.1016/S0044-8486(03)00501-5.

[68] Alver MO, Alfredsen JA, Øie G, Storøy W, Olsen Y. Automatic control of growth and density in rotifer cultures. Aquac Eng 2010;43:6-13. https://doi.org/10.1016/j.aquaeng.2010.02.003.

[69] Han J, Lee K-W. Influence of salinity on population growth, oxidative stress and antioxidant defense system in the marine monogonont rotifer *Brachionus plicatilis*. Comp Biochem Physiol Part B Biochem Mol Biol 2020;250:110487. https://doi.org/10.1016/j.cbpb.2020.110487.

[70] Chilmawati D, Suminto. The Effect of Different Diet of Phytoplankton Cells on Growth Performance of Copepod, *Oithona* sp. in Semi-mass Culture. Aquat Procedia 2016;7:39-45. https://doi.org/10.1016/j.aqpro.2016.07.005.

[71] Knuckey RM, Semmens GL, Mayer RJ, Rimmer MA. Development of an optimal microalgal diet for the culture of the calanoid copepod *Acartia sinjiensis*: Effect of algal species and feed concentration on copepod development. Aquaculture 2005;249:339-351. https://doi.org/10.1016/j.aquaculture.2005.02.053.

[72] Puello-Cruz AC, Mezo-Villalobos S, González-Rodríguez B, Voltolina D.

Culture of the calanoid copepod *Pseudodiaptomus euryhalinus* (Johnson 1939) with different microalgal diets. Aquaculture 2009;290:317-319. https://doi.org/10.1016/j.aquaculture.2009.02.016.

[73] Ribeiro ACB, Souza-Santos LP. Mass culture and offspring production of marine harpacticoid copepod *Tisbe biminiensis*. Aquaculture 2011;321:280-288. https://doi.org/10.1016/j.aquaculture.2011.09.016.

[74] Sarkisian BL, Lemus JT, Apeitos A, Blaylock RB, Saillant EA. An intensive, large-scale batch culture system to produce the calanoid copepod, *Acartia tonsa*. Aquaculture 2019;501:272-278. https://doi.org/10.1016/j.aquaculture.2018.11.042.

[75] Liu Z, Yu P, Cai M, Wu D, Zhang M, Huang Y, et al. Polystyrene nanoplastic exposure induces immobilization, reproduction, and stress defense in the freshwater cladoceran *Daphnia pulex*. Chemosphere 2019;215:74-81. https://doi.org/10.1016/j.chemosphere.2018.09.176.

[76] Raymundo LB, Rocha O, Moreira RA, Miguel M, Daam MA. Sensitivity of tropical cladocerans to chlorpyrifos and other insecticides as compared to their temperate counterparts. Chemosphere 2019;220:937-942. https://doi.org/10.1016/j.chemosphere.2019.01.005.

[77] Jaikumar G, Baas J, Brun NR, Vijver MG, Bosker T. Acute sensitivity of three Cladoceran species to different types of microplastics in combination with thermal stress. Environ Pollut 2018;239:733-740. https://doi.org/10.1016/j.envpol.2018.04.069.

[78] Hansen B. Aspects of feeding, growth and stage development by trochophora larvae of the boreal polychaete *Mediomastus fragile* (Rasmussen) (Capitellidae). J Exp Mar Bio Ecol 1993;166:273-288. https://doi.org/10.1016/0022-0981(93)90224-C.

[79] Pulz O, Gross W. Valuable products from biotechnology of microalgae. Appl Microbiol Biotechnol 2004;65:635-648. https://doi.org/10.1007/s00253-004-1647-x.

[80] Nethravathy MU, Mehar JG, Mudliar SN, Shekh AY. Recent Advances in Microalgal Bioactives for Food, Feed, and Healthcare Products: Commercial Potential, Market Space, and Sustainability. Compr Rev Food Sci Food Saf 2019;18:1882-1897. https://doi.org/10.1111/1541-4337.12500.

[81] Nagaraj S, Rajaram MG, Arulmurugan P, Baskaraboopathy A, Karuppasamy K, Jayappriyan KR, et al. Antiproliferative potential of astaxanthin-rich alga *Haematococcus pluvialis* Flotow on human hepatic cancer (HepG2) cell line. Biomed Prev Nutr 2012;2:149-153. https://doi.org/10.1016/j.bionut.2012.03.009.

[82] Parodi A, Leip A, Slegers PM, Ziegler F, Herrero M, Tuomisto H, et al. Future foods : towards a sustainable and healthy diet for a growing population. Nat Sustain 2018;1:782-789.

[83] Derner RB, Ohse S, Villela M, Carvalho SM de, Fett R. Microalgas, produtos e aplicações. Ciência Rural 2006;36:1959-1967. https://doi.org/10.1590/S0103-847820060 00600050.

[84] Ferreira SP, Souza-Soares L, Costa JA. Revisão: microalgas: uma fonte alternativa na obtenção de ácidos gordos essenciais. Rev Ciências Agrárias 2013;36:275-328.

[85] Mishra N, Gupta E, Singh P, Prasad R. Application of microalgae metabolites in food and pharmaceutical industry. In: Egbuna C, Mishra A, Goyal M, editors. Prep. Phytopharm. Manag. Disord., Cambridge, MA, USA:

Academic Press; 2020, p. 391-408. https://doi.org/https://doi.org/10.1016/B978-0-12-820284-5.00005-8.

[86] Lafarga T, Fernández-Sevilla JM, González-López C, Acién-Fernández FG. Spirulina for the food and functional food industries. Food Res Int 2020;137:109356. https://doi.org/10.1016/j.foodres.2020.109356.

[87] Koukouraki P, Tsoupras A, Sotiroudis G, Demopoulos CA, Sotiroudis TG. Antithrombotic properties of Spirulina extracts against platelet-activating factor and thrombin. Food Biosci 2020;37:100686. https://doi.org/10.1016/j.fbio.2020.100686.

[88] Becker W. Microalgae in Human and Animal Nutrition. In: Richmond A, editor. Handb. Microalgal Cult., Oxford, UK: Blackwell Publishing Ltd; 2004, p. 312-51. https://doi.org/10.1002/9780470995280.ch18.

[89] Liang S, Liu X, Chen F, Chen Z. Current microalgal health food R & D activities in China. In: Ang PO, editor. Asian Pacific Phycol. 21st Century Prospect. Challenges, Dordrecht, Netherlands: Springer Netherlands; 2004, p. 45-8. https://doi.org/10.1007/978-94-007-0944-7_7.

[90] Kirchmayr A, Griesbeck C. 5 Genetic engineering, methods and targets. In: Posten C, Walter C, editors. Microalgal Biotechnol. Potential Prod., Berlin, Germany: De Gruyter; 2012. https://doi.org/10.1515/9783110225020.87.

[91] Sirakov I, Velichkova K, Stoyanova S, Staykov Y. The importance of microalgae for aquaculture industry. Review. Int J Fish Aquat Stud 2015;2:81-84.

[92] Spolaore P, Joannis-Cassan C, Duran E, Isambert A. Commercial applications of microalgae. J Biosci Bioeng 2006;101:87-96. https://doi.org/10.1263/jbb.101.87.

[93] Madeira MS, Cardoso C, Lopes PA, Coelho D, Afonso C, Bandarra NM, et al. Microalgae as feed ingredients for livestock production and meat quality: A review. Livest Sci 2017;205:111-121. https://doi.org/10.1016/j.livsci.2017.09.020.

[94] Grubišić M, Ivančić Šantek M, Šantek B. Potential of microalgae for the production of different biotechnological products. Chem Biochem Eng Q 2019;33:161-181. https://doi.org/10.15255/CABEQ.2019.1657.

[95] Rawat, I.; Gupta, S.K.; Shriwastav, A.; Singh, P.; Kumari, S.; Bux F. Microalgae Applications in Wastewater Treatment. Algae Biotechnol 2016:249-268.

[96] Quijano, G.; Arcila, J.S.; Buitrón G. Microalgal-bacterial aggregates: Applications and perspectives for wastewater treatment. Biotechnol Adv 2017;35:772-781.

[97] Satyanarayana, K.G.; Mariano, A.B.; Vargas JVC. A review on microalgae, a versatile source for sustainable energy and materials. Int J Energy Res 2011;35:291-311.

[98] Ravindran, B.; Gupta, S.K.; Cho, W.M.; Kim, J.K.; Lee, S.R.; Jeong, K.H.; Lee, D.J.; Choi HC. Microalgae potential and multiple roles-current progress and future prospects-an overview. Sustainability 2016;8:1215.

[99] Emparan Q, Harun QR, DAnquah MK. Role of phycoremediation for nutrient removal from wastewaters: a review 2019;17:889-915.

[100] Agu A, Garcı JF, Molina-dı A. Chemical evaluation of contaminants in wastewater effluents and the environmental risk of reusing effluents in agriculture 2009;28. https://doi.org/10.1016/j.trac.2009.03.007.

[101] Menció A, Mas-pla J, Otero N, Regàs O, Boy-roura M, Puig R, et al.

Nitrate pollution of groundwater; all right ..., but nothing else ? Sci Total Environ 2016;539:241-251. https://doi.org/10.1016/j.scitotenv.2015.08.151.

[102] Chowdhury S, Mazumder MAJ, Al-attas O, Husain T. Science of the Total Environment Heavy metals in drinking water : Occurrences, implications, and future needs in developing countries. Sci Total Environ 2016;569-570:476-488. https://doi.org/10.1016/j.scitotenv.2016.06.166.

[103] Richards, R.G.; Mullins BJ. Using Microalgae for Combined Lipid Production and Heavy Metal Removal from Leachate. Ecol Modell 2013;249:59-57.

[104] Chiaiese, P.; Palomba, F.; Tatino, F.; Lanzillo, C.; Pinto, G.; Pollio, A.; Filippone E. Engineered Tobacco and Microalgae Secreting the Fungal Laccase POXA1b Reduce Phenol Content in Olive Oil Mill Wastewater. Enzyme Microb Technol 2011;49:540-546.

[105] Marques AE, Miranda JR, Batista AP, Gouveia L. Microalgae Biotechnological Applications: Nutrition, Health and Environment. In: Johansen MN, editor. Microalgae Biotechnol. Microbiol. Energy, Hauppauge, NY, USA: Nova Science Publishers, Inc.; 2013, p. 1-59.

[106] Rodolfi L, Chini Zittelli G, Bassi N, Padovani G, Biondi N, Bonini G, et al. Microalgae for oil: Strain selection, induction of lipid synthesis and outdoor mass cultivation in a low-cost photobioreactor. Biotechnol Bioeng 2009;102:100-112. https://doi.org/10.1002/bit.22033.

[107] Harun R, Yip JWS, Thiruvenkadam S, Ghani WAWAK, Cherrington T, Danquah MK. Algal biomass conversion to bioethanol - a step-by-step assessment. Biotechnol J 2014;9:73-86. https://doi.org/10.1002/biot.201200353.

Section 2

Phytoplankton

Chapter 3

Remote Sensing of Phytoplankton Pigments

Guoqing Wang and John Moisan

Abstract

Pigments, as a vital part of phytoplankton, act as the light harvesters and protectors in the process of photosynthesis. Historically, most of the previous studies have been focused on chlorophyll *a*, the primary light harvesting pigment. With the advances in technologies, especially High-Performance Liquid Chromatography (HPLC) and satellite ocean color remote sensing, recent studies promote the importance of the phytoplankton accessory pigments. In this chapter, we will overview the technology advances in phytoplankton pigment identification, the history of ocean color remote sensing and its application in retrieving phytoplankton pigments, and the existing challenges and opportunities for future studies in this field.

Keywords: phytoplankton, pigments, remote sensing, ocean color, satellite

1. Introduction

Phytoplankton live near the water surface to capture sufficient light for photosynthesis and act as the primary producer of the plankton community. They form the bottom levels of the marine and aquatic food webs, and their existence not only makes life in the water possible but also makes the ocean an important food source for mankind. Phytoplankton play a crucial role in the biogeochemical cycles of many important chemical elements, not only carbon but also of other elements, such as silica and nitrogen [1–4]. The release and uptake of CO_2 and CH_4, and the excretion of dimethylsulphide by phytoplankton influence the atmosphere and climate [5]. As a result of the changes in their living condition, their composition and concentration vary over space and time, which in turn can influence the whole ecosystem, such as through the changes in the size structure, formation of harmful algal blooms and development of hypoxic regions. Blooms and hypoxia can disrupt food-webs and threaten human health.

Phytoplankton pigments capture sunlight. The resulting photosynthesis and its products, especially the oxygen and organic compounds, all rely on the light energy captured by the different phytoplankton pigments [6–8]. Chlorophyll *a* is the major pigment for light harvesting. Accessory pigments (e.g. chlorophylls *b* and *c*, carotenoids, and phycobiliproteins) also play a significant role in photosynthesis and photoprotection, by extending the light collection window and protecting the cell from damage of high irradiance levels or high ultraviolet light exposure. With the commercial availability of fluorometers, routine measurements of chlorophyll *a* became possible. That single technology to measure chlorophyll *a* fluorescence made the measurement a universal parameter for estimating phytoplankton biomass and productivity. As a result of improvements in culturing, microscopy, HPLC and molecular methods, rapidly separating and quantifying pigments from different phytoplankton has become possible [9–11]. These new measurements make it possible to use phytoplankton pigments as indicators to elucidate the composition and fate of phytoplankton in the world's oceans [12].

Light absorbed by phytoplankton pigments provides the initial energy for carbon cycles, and is also one of the major factors influencing the appearance of water color [13–16]. To study this important water column phenomenon, ocean color remote sensing was first proposed in late 1970s. Satellite-based ocean color remote sensing provides unique observational capability to scientists for phytoplankton studies by providing synoptic views of the ocean with high spatial and temporal resolution. Since the Coastal Zone Color Scanner (CZCS) mission, chlorophyll *a* retrieval has been the principle focus of ocean color remote sensing research (e.g., [17]). Whereas this focus continues to the present [18–20], an evolving interest in retrieving other pigments, has emerged in recent years.

What follows, based on the most recent research findings from the ocean color community, is a brief review of how phytoplankton pigments are estimated from water samples, how pigment maps are derived from satellite measurements and what are the existing challenges and opportunities for the estimates and application of remote sensed pigments. This chapter is not meant to present a comprehensive list of all possible topics related to satellite-based pigment observations, but rather its focus is on the history of pigment retrievals with several examples showing major findings. For interested readers, a full breadth and depth knowledge in this field can be obtained by reading the refereed literature and technical reports compiled on the National Aeronautics and Space Administration ocean color website (https://oceancolor.gsfc.nasa.gov) and by International Ocean Color Coordinating Group (http://www.ioccg.org).

2. Phytoplankton and pigment properties

2.1 Optical properties

2.1.1 Absorption properties

Optical properties of phytoplankton, especially the absorption coefficients of the pigments inside them (**Figure 1**), play a key role in determining not only the use of this radiant energy for photosynthesis, but also the penetration of the radiant energy within water. These pigment absorption coefficients are important for identifying and quantifying phytoplankton groups [12] and size class distributions (IOCCG report 15 and references therein), understanding of photosynthetic rate [11, 21], and in particular for ocean color interpretation.

Light absorption properties of phytoplankton cells from laboratory cultures as experimental materials have received a great deal of attention in fundamental photosynthesis research [22, 23]. However, the phytoplankton pigment absorption properties from natural water is the information needed in ocean color remote sensing. The collection of phytoplankton pigment information has been obtained from measurement of the spectral absorption of phytoplankton, usually through filtration onto a filter pad because of the low *in situ* concentrations of phytoplankton in the water [24].

Using data on pigment concentrations and their absorption properties, Kirkpatrick *et al.* [25] used the specific pigment absorption peaks for identification of phytoplankton types. This method has been integrated into spectral shape-based

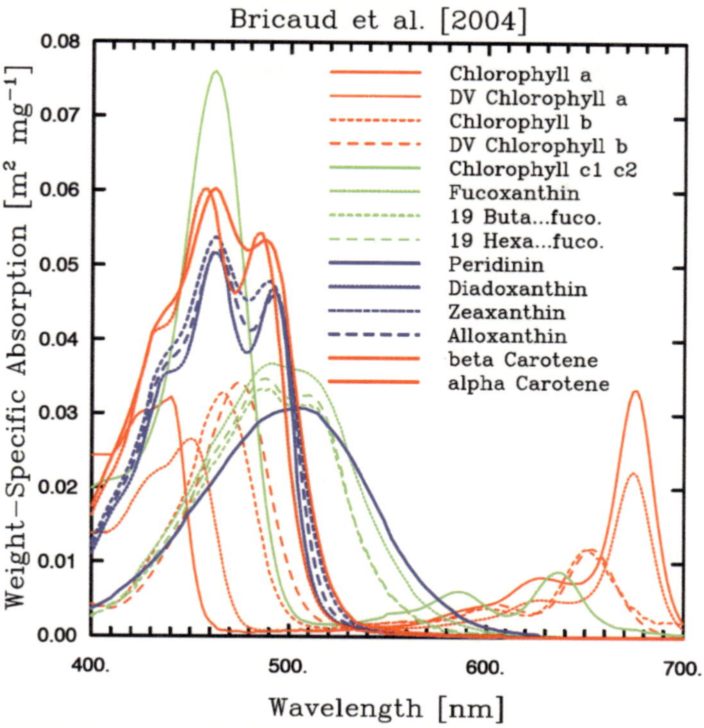

Figure 1.
Weight-specific (or pigment-specific) in vitro absorption spectra of various pigments derived from measuring the absorption spectra of individual pigments in solvent and shifting the maxima of the spectra according to Bidigare et al. [14]. Data obtained courtesy of Annick Bricaud (See Bricaud et al. [15]). Credit to Moisan et al. [30].

remote sensing algorithms [26, 27]. However, the absorption of phytoplankton is more complicated than a simple sum of the absorption properties of individual pigments. Differences in pigment composition and the pigment package effect influence not only the magnitude but also the shape of the spectra of phytoplankton absorption [14, 15, 28–30]. All these introduce variabilities in the specific absorption coefficients and increase the uncertainties in the application of such information.

Hoepffner and Sathyendranath [29] proposed Gaussian decomposition of phytoplankton absorption spectra. For the first time, this method decomposed the absorption spectra into Gaussian curve components and linked them to the light absorption coefficients of multiple pigments inside phytoplankton cells. Several studies followed this proxy to estimate multiple phytoplankton pigments for different water bodies [31–33] but were limited to using only *in situ* measured absorption coefficients. Wang et al. [34, 35] proposed a semi-analytical algorithm to obtain these Gaussian curves and pigment absorption coefficients from ocean color remote sensing data.

2.1.2 Fluorescence

A portion of the light absorbed by phytoplankton pigments can be emitted at a longer wavelength in a physical process called fluorescence [36]. The energy dissipated in fluorescence is secondary to the amount absorbed and used for photosynthesis, but it is still significant enough to be observed in ocean color remote sensing data. Chlorophyll *a* fluorescence has been the most significantly used fluorescence feature (**Figure 2**), and the detection and products from satellite ocean color sensors have been widely used [37, 38]. Several other phytoplankton pigments (pheopigments and phycobilins) can also fluoresce.

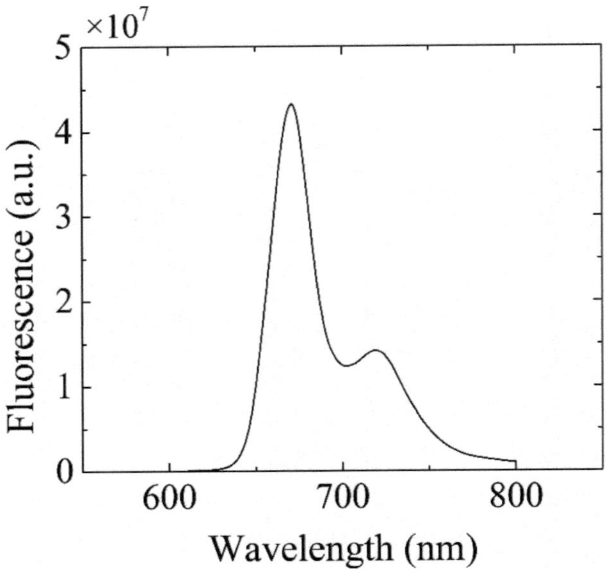

Figure 2.
Chlorophyll a *fluorescence emission. Data from Du et al. [42] and Dixon et al. [43].*

Several factors influence phytoplankton fluorescence: nutrient conditions, stage of growth, physiological state of phytoplankton, pigment content and ratios, taxonomic position of algae, and photoadaptation [39–41]. *In situ* chlorophyll fluorescence has been the most frequent method for describing the chlorophyll and phytoplankton variation and distribution in the ocean [41], but all the uncertainties from the pigment properties make the interpretation of the chlorophyll fluorescence data a challenge.

2.2 Pigment measurements

Historically, chlorophyll *a* has been routinely derived from filtered fluorometric measurements following standard methods using commercially availability of fluorometers. However, even standard methods yield varying results depending on the composition of pigments within the phytoplankton, and errors can be on the order of 50% [44–46]. The presence of significant amount of chlorophyll *b* and/or chlorophyll *c*, causes fluorometric techniques to under- or over-estimate Chlorophyll *a* with respect to fluorometric measurements [44–47]. The pigment package effect is also a major source of concern.

The introduction of pigment analyses by high-pressure liquid chromatography (HPLC) [48, 49] facilitated easy and accurate separation, identification, and quantification of phytoplankton pigments. Pigment detection based on HPLC methods enables quantification of over 50 phytoplankton pigments [11, 50]. Some of the pigments can be used as diagnostic pigments for phytoplankton groups (e.g., fucoxanthin for diatoms, peridinin for dinoflagellates, alloxanthin for cryptophytes, chlorophyll *b* for chlorophytes, 19′-hex-fucoxanthin for haptophytes, and 19′-but-fucoxanthin for pelagophytes) [51, 52]. Moreover, diadinoxanthin and diatoxanthin are generally found in dinoflagellates (Phylum Miozoa, Class Dinophyceae) and diatoms (Phylum Bacillariophyta, Class Bacillariophyceae), whereas lutein, prasinoxanthin, neoxanthin, and violaxanthin are found in class Chlorophyceae (Phylum Chlorophyta) and class Prasinophyceae (Phylum Chlorophyta). Chlorophyll *a*, *c*, and β-carotene are used as general indicators of total algal biomass. Phytoplankton are also often categorized into three different groups: micro-phytoplankton (20–200 μm), nano-phytoplankton (2–20 μm), and pico-phytoplankton (0.2–2 μm) [53]. The contribution of each group can also be calculated using its pigment signatures [54].

3. Ocean color remote sensing

Ocean color or aquatic remote sensing refers to the use of optical measurements made from aircraft or satellites to obtain information about the constituents of the waters.

Remote sensing can be classified as active or passive based on the energy source. Active remote sensing shots signal from the sensor platform (satellite or aircraft) to the water body and detects the return signal from it. Passive remote sensing observes the light that is reflected or emitted by the water body. The most commonly used light source for passive remote sensing is sunlight. Sensors detect the reflected or backscattered light coming from the water body. The launch of the first ocean color sensor Coastal Zone Color Scanner (CZCS) in 1978, started the era for passive satellite ocean color remote sensing.

Passive ocean-color remote sensing is conceptually simple (**Figure 3**). The signals captured by remote sensors provide information on the types and concentrations of the various constituents of the water body. The concentrations of optically-active substances present in the water can be estimated by inverting bio-optical algorithms with remote sensing data. Although this process can be fraught with difficulties, our understanding of the oceans has been completely revolutionized by ocean color remote sensing from daily to decadal temporal scales and local to global spatial scales

For a better understanding of phytoplankton in the global ocean from large spatial and temporal scales, ocean color remote sensing is the most efficient tool, with the advantages of cost-free satellite imagery access from NASA and others,

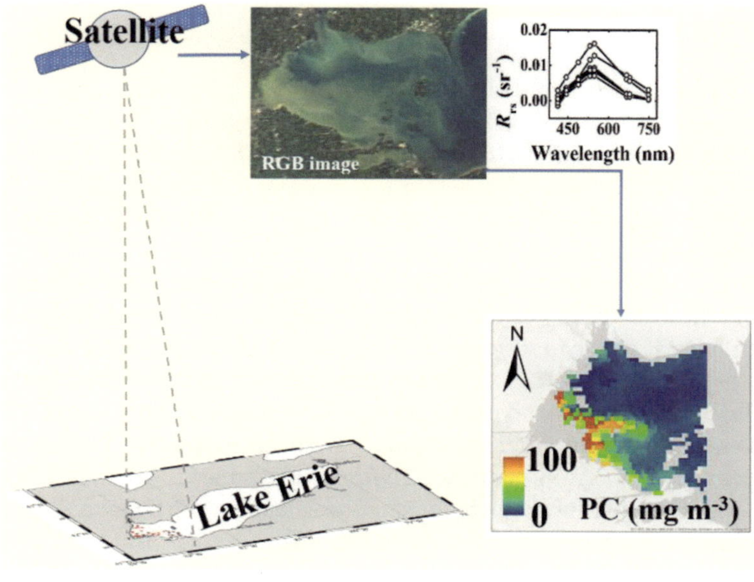

Figure 3.
Conceptual figure of passive satellite ocean color remote sensing with Western Lake Erie as an example: $R_{rs}(\lambda)$ *as remote sensing reflectance, PC: pigment concentration.*

thus providing a data source for hypothesis testing and more efficient utilization of limited *in situ* data.

Phytoplankton pigments have a major effect on ocean color and are one of the primary reasons for studying it. Following the launch of CZCS, unprecedented data for studying the biology of the oceans have been obtained [55]. For the first time, chlorophyll *a* concentration in the surface ocean could be estimated at synoptic scales [56, 57], leading to unprecedented understanding of the biogeochemistry of the ocean, e.g., primary productivity [58]. These ocean-color observations were continued by the Sea-viewing Wide Field-of-view Sensor (SeaWiFS) mission in 1997, which was then followed by the Moderate Resolution Imaging Spectroradiometer (MODIS on Terra in 2000, and Aqua in 2002), the Medium Resolution Imaging Spectrometer (MERIS, 2002–2012), the Visible Infrared Imaging Radiometer Suite (VIIRS, 2011 – present), and the upcoming hyperspectral Plankton, Aerosol, Cloud, ocean Ecosystem (PACE) mission (planned to launch in 2023).

3.1 Remote sensing of pigments

In the past decades, the identification of phytoplankton pigments from satellite remote sensing has been mainly focused on chlorophyll *a*, and the products have been widely used to represent the phytoplankton biomass in the primary productivity estimation and biogeochemical models. With the increasing recognition of the important role accessory pigments play, remote sensing of pigments from space form this rapidly advancing field. High temporal and spatial monitoring are particularly important for the study of harmful algal blooms (HABs, e.g. cyanobacteria, [59, 60]). These blooms are often toxic and a growing problem in many coastal and inland waters of the world. A review of chlorophyll *a* algorithm for global oceans has been provided in recent papers including Dierssen [61] and Hu and Campbell [62]. In general, the method to obtain phytoplankton pigments from satellite remote sensing can be classified into two different categories: empirical, and semi-analytical.

3.1.1 Empirical methods

In the process of obtaining phytoplankton pigment, especially chlorophyll *a* (Chl-a) concentrations, most effort has focused on empirical algorithms, not only because of the simplicity, but also the effectiveness. The empirical methods estimate pigments from satellite derived remote sensing reflectance ($R_{rs}(\lambda)$) through regression of pigment concentrations against $R_{rs}(\lambda)$ band ratios or band differences (e.g., [20, 63, 64]).

These methods account for regional variabilities in water properties and $R_{rs}(\lambda)$ input errors through tuning of the empirical coefficients, although the empirical design makes it prone to influences from various in-water constituents. The spectrally dependent $R_{rs}(\lambda)$ errors [65] to a large extent could be compensated through the band ratio or band difference used in empirical approaches. Thus, from the CZCS era, a set of empirical algorithms have been adopted by U.S. National Aeronautics and Space Administration (NASA) to produce the default Chl-a products from the existing ocean color satellite sensors, even though these empirical Chl-a products contain large uncertainties [61, 66].

For remote sensing of accessory pigments, Pan *et al.* [67] proposed to retrieve 17 different phytoplankton pigments from satellite remote sensing data using empirical methods and applied the information to phytoplankton group identification

[68]. This method simply used empirical relationships between pigment concentrations with the ratio of two remote sensing reflectance bands (488 or 490 to 547 or 555 nm). However, same as Chl-a, in optically complicated coastal and inland waters, higher uncertainties could be introduced by the large influences from colored detrital matters (CDM) in coastal waters.

Eq. (1) shows the polynomial algorithm for pigments, in which the blue-green band ratio was empirically related to pigment concentrations (C_{pigs}):

$$\log_{10}\left(C_{pigs}\right) = a_0 + \sum_{i=1}^{N} a_i \left(\log_{10}\left(\frac{R_{rs}(\lambda_1)}{R_{rs}(\lambda_2)} \right) \right)^i \tag{1}$$

Where λ_1 and λ_2 represent the spectral band around blue (440–520) and green (555) region respectively, and $a_0 - a_N$ are sensor specific regression coefficients. Details of the spectral bands and parameters used for each sensor can be found in [67] and on NASA ocean color website for Chl-a: https://oceancolor.gsfc.nasa.gov/atbd/chlor_a/.

3.1.2 Semi-analytical algorithms

The semi-analytical algorithms obtain pigments from $R_{rs}(\lambda)$ by solving a series of equations established from simplified radiative transfer theory based on several bio-optical assumptions (e.g., [69–73]). In principle, these methods have the potential to obtain more accurate results than the empirical methods because the different water constituents affecting water color are explicitly separated. However, semi-analytical approach has its own strengths and weaknesses. Semi-analytical methods rely on tuning of the empirical parameters in the bio-optical relationships using global or local datasets. As a result of the optical properties of the constituents, the separation of them from $R_{rs}(\lambda)$ is not as explicit as expected.

Semi-analytical algorithms are relatively more complex. Based on the radiative transfer equation, remote sensing reflectance was defined as the ratio of upwelling radiance to downwelling irradiance, and its relationship with inherent optical properties of water constituents can be expressed as:

$$R_{rs}(\lambda) = G \frac{b_{bw}(\lambda) + b_{bp}(\lambda)}{a_w(\lambda) + a_{ph}(\lambda) + a_{CDOM}(\lambda) + a_{NAP}(\lambda) + b_{bw}(\lambda) + b_{bp}(\lambda)} \tag{2}$$

Where G is a parameter related to the environment and solar and sensor viewing geometry. The absorption coefficients of water ($a_w(\lambda)$), phytoplankton ($a_{ph}(\lambda)$), colored dissolved organic matter ($a_{CDOM}(\lambda)$), non-algal particles ($a_{NAP}(\lambda)$), and backscattering coefficients of water ($b_{bw}(\lambda)$) and particles ($b_{bp}(\lambda)$).

Pigment concentrations can be estimated from phytoplankton absorption coefficients from Gaussian decomposition (Eqs. 3 and 4) or by using pigment specific absorption coefficients (Eq. 5). **Figure 4** shows an example of Chl-a global distribution map obtained from MERIS ocean color data using a semi-analytical algorithm.

$$a_{ph}(\lambda) = \sum_{i=1}^{n} a_{Gau}(\lambda_i) \exp\left[-0.5 \left(\frac{\lambda - \lambda_i}{\sigma_i} \right)^2 \right] \tag{3}$$

$$\log_{10}\left(C_{pigs}\right) = a_0 + \sum_{i=1}^{n} a_i \log_{10}\left(a_{Gau}(\lambda_i)\right) \tag{4}$$

where σ_i and $a_{Gau}(\lambda_i)$ are the width and peak magnitude of the ith Gaussian curve at peak center (λ_i). As shown in **Figure 1**, in the Gaussian curve assumption in Hoepffner and Sathyendranath [29], each Gaussian curve represents the absorption curve of a specific pigment. C_{pigs} are pigment concentrations, with a_0 and a_i as empirical parameters [74].

$$a_{ph}(\lambda) = \sum_{i=1}^{N} C_{pigi} a^{*}_{pigi} \tag{5}$$

With a^{*}_{pig} as the pigment specific absorption coefficients [14, 15, 75, 76].

3.2 Application of remote sensed pigments

The measuring of ocean color from space and the increasing accuracy of *in situ* pigment measurements for determining phytoplankton groups and types in the water column have greatly facilitated progress in phytoplankton research.

Empirical algorithms used to calculate chlorophyll *a* concentration from ocean color data were established for different waters (*e.g.*, [17, 19, 60, 63, 77–79]). The development and application of spectral inversion algorithms to ocean color data have further provided assessments of absorption by phytoplankton pigment [34, 71, 72, 80–83]. Additional algorithm development using these properties has led to new retrievals regarding plankton community composition, including phytoplankton size fractions, the slope of the particle size distribution, and even specific phytoplankton groups, such as coccolithophores (Phylum Haptophyta, Class Coccolithophyceae), Trichodesmium (Phylum Cyanobacteria), and harmful algal species (e.g., [84–99] and references therein).

In recent years, the use of pigment data to map phytoplankton population and composition in the water column has become an established and convenient way of studying field phytoplankton [100]. Phytoplankton biomass and the structure of phytoplankton community have been widely quantified and assessed using photosynthetic pigment biomarkers [52, 100]. Photosynthetic pigments also function as indicators of the physiological condition of a phytoplankton community, which may be affected by environmental and trophic conditions [101]. Photosynthetic carotenoids (PSC) are dominant in high productivity waters,

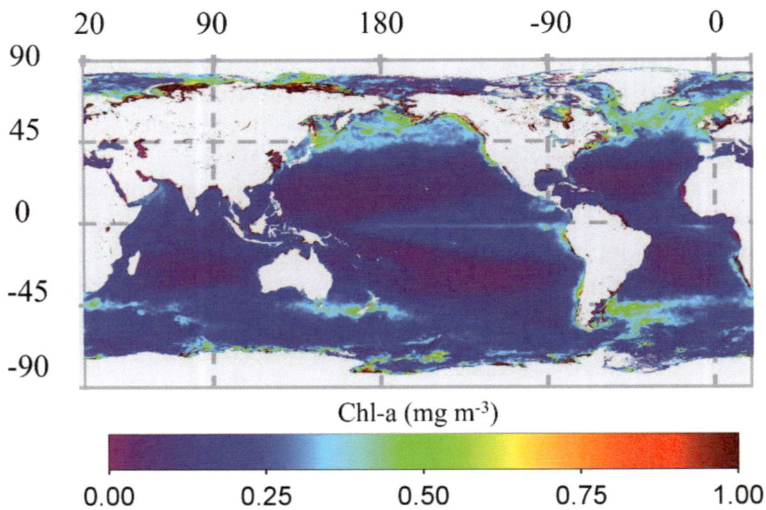

Figure 4.
Chlorophyll a map of the global ocean from MERIS for the year of 2007 with data from Wang et al. [74].

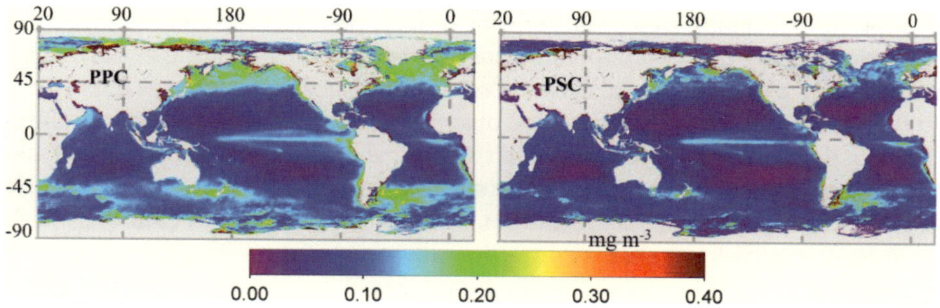

Figure 5.
Global maps of photoprotective (PPC) and photosynthetic carotenoids (PSC) from Wang et al. [74].

whereas photoprotective carotenoids (PPC) are more dominant in low productivity waters [102, 103]. In addition, intensive light increases the PPC:PSC ratio [104, 105]. Thus, the PPC:PSC ratio can be used as a good indicator of changes in environmental factors. **Figure 5** shows the global maps of PPC and PSC from Wang et al. [74].

The sustained time series of these phytoplankton properties from ocean color remote sensing has provided major advances in our understanding of carbon dynamics, plankton annual cycles and their responses to climate variations. Simply, the satellite ocean color remote sensing of pigment will further improve the research revolution in oceanography.

4. Challenges and opportunities

4.1 Uncertainties in satellite remote sensing data

Although ocean color remote sensing observations enabled advances in our understanding of phytoplankton in the ocean, there are several fundamental limitations in the passive radiometric technique. The major uncertainties of remote sensing pigment estimates are from atmospheric correction errors, as a result of the high signal contribution of components other than the targeted water to radiances measured by ocean color instruments, such as reflection from the ocean surface, surface foam, subsurface bubbles, and atmospheric constituents, including clouds, aerosols, and air molecules. A small error from the correction of these atmospheric contribution results in large errors in the obtained remote sensing reflectance and the associated pigment information ([106] and references therein).

Another challenge with ocean color remote sensing comes from the interferences of the optical properties of retrieved water components, including absorption by phytoplankton pigments, colored dissolved matter, and nonalgal particles, and backscattering by suspended particles. This makes the uncertainties from these properties and the derived geophysical parameters from them hard to reduce. The upcoming PACE mission is designed with expanded spectral range and resolution to address this problem [107].

Finally, clouds and strongly scattering aerosol layers have been significant limitation factors of the availability of satellite ocean color data. On average, about 70% of the Earth's ocean area were covered by clouds on the daily scene obtained from a sensor. For broken cloud or aerosol interfered scenes, the accuracy of ocean color retrievals can be compromised compared to clear sky pixels. In high altitude regions, specifically the polar regions, cloud conditions and low sun angles limited ocean color sampling from late fall through early spring of next year. The lack of sampling for this long period of time makes it impossible for a complete understanding of the biogeochemistry and plankton annual cycles of some of the most productive waters [108].

Other issues are from the limitation of spectral, spatial, and temporal resolutions of the existing satellite sensors: some harmful algal blooms occurring in small lakes and ponds are not able to be detected by satellite sensors with low spatial resolution (~1 km); while the high spatial resolution sensors (e.g., Landsat 8) cannot provide timely coverage of bloom events due to their low temporal resolution.

4.2 More accurate *in situ* measurements

The satellite ocean color remote sensing has been tasked to acquire remote sensing imagery, validate and monitor its accuracy, process the radiometric data into geophysical information using different algorithms, and apply the final products into scientific research. One of the principles of *in situ* datasets for the calibration and validation procedure is estimates of near-surface pigment concentrations [109]. Thus, accurate and complete pigment measurements are important to algorithm development as used with remote sensing of phytoplankton pigments. The application of pigment chemotaxonomy in oceanography will be more firmly established by advances in taxonomy and improved pigment analysis (e.g. greater resolution with advanced HPLC and ultra-high performance liquid chromatography – UPLC), more rapid and secure chemical identification, and further measurement and estimation of in vivo pigment absorption coefficients. With improvement in these techniques, more discoveries in pigment and taxonomic diversity and further understanding of their influences on the biogeochemical cycles of the ocean will be achieved. The current challenging environment from climate change makes this an urgent need [14, 15, 75, 76, 91, 110, 111].

4.3 Active remote sensing: LIDAR

Compared to passive ocean color remote sensing, lidar shows many advantages, such as operating at night and high latitudes, and can generally penetrate to the subsurface chlorophyll maximum [112, 113]. Airborne lidar is particularly useful for mapping the depth distribution of phytoplankton. The characteristic depth profiles of phytoplankton provide useful information for differentiation of phytoplankton species as described in Moore et al. [114] two different species of harmful Cyanobacteria in Lake Erie, USA can be identified by the differences in their characteristic depth profiles.

Combining the observations from lidar and ocean color sensors, especially the advanced upcoming PACE mission, would enable the achievement of greater synergies. The pairing of an ocean-optimized satellite profiling lidar with a passive ocean color sensor would provide maximized global data coverage, and enable three-dimensional reconstruction of ocean ecosystems, which would further favor the algorithm development, and expand the retrieval of geophysical properties.

Acknowledgements

We thank the National Aeronautics and Space Administration (NASA) for providing the MERIS imagery, and the support from the NASA Advanced Information Systems Technology (AIST) program.

DOI: http://dx.doi.org/10.5772/intechopen.95381

Author details

Guoqing Wang[1,2]* and John Moisan[3]

1 NASA Goddard Space Flight Center, Greenbelt, MD, USA

2 Science Systems and Applications, Inc., MD, USA

3 NASA Goddard Space Flight Center, Wallops Flight Facility, Wallops Island, VA, USA

*Address all correspondence to: guoqing.wang@ssaihq.com

References

[1] Nelson, D. M., & Smith Jr, W. O. (1986). Phytoplankton bloom dynamics of the western Ross Sea ice edge—II. Mesoscale cycling of nitrogen and silicon. Deep Sea Research Part A. Oceanographic Research Papers, 33(10), 1389-1412.

[2] Tréguer, P., & Jacques, G. (1992). Review Dynamics of nutrients and phytoplankton, and fluxes of carbon, nitrogen and silicon in the Antarctic Ocean. In Weddell Sea Ecology (pp. 149-162). Springer, Berlin, Heidelberg.

[3] Van Bennekom, A. J., Berger, G. W., Van der Gaast, S. J., & De Vries, R. T. P. (1988). Primary productivity and the silica cycle in the Southern Ocean (Atlantic sector). Palaeogeography, Palaeoclimatology, Palaeoecology, 67 (1-2), 19-30.

[4] Dortch, Q., & Whitledge, T. E. (1992). Does nitrogen or silicon limit phytoplankton production in the Mississippi River plume and nearby regions. Continental Shelf Research, 12(11), 1293-1309.

[5] Charlson, R. J., Lovelock, J. E., Andreae, M. O., & Warren, S. G. (1987). Oceanic phytoplankton, atmospheric sulphur, cloud albedo and climate. Nature, 326(6114), 655-661.

[6] Kirk, J. T. (1994). Light and photosynthesis in aquatic ecosystems. Cambridge university press.

[7] Behrenfeld, M. J., & Falkowski, P. G. (1997). Photosynthetic rates derived from satellite-based chlorophyll concentration. Limnology and oceanography, 42(1), 1-20.

[8] Falkowski, P. G. (2002). The ocean's invisible forest. Scientific American, 287(2), 54-61.

[9] Wright, S. W., Jeffrey, S. W., Mantoura, R. F. C., Llewellyn, C. A., Bjørnland, T., Repeta, D., & Welschmeyer, N. (1991). Improved HPLC method for the analysis of chlorophylls and carotenoids from marine phytoplankton. Marine ecology progress series, 183-196.

[10] Roy, S., Llewellyn, C. A., Egeland, E. S., & Johnsen, G. (Eds.). (2011). Phytoplankton pigments: characterization, chemotaxonomy and applications in oceanography. Cambridge University Press.

[11] Jeffrey, S.W.; Mantoura, R.F.C.; Wright, S.W. (Ed.) (1997). Phytoplankton pigments in oceanography: guidelines to modern methods. Monographs on Oceanographic Methodology, 10. UNESCO Publishing: Paris. ISBN 92-3-103275-5. 661 pp.

[12] Schlüter, L., Møhlenberg, F., Havskum, H., & Larsen, S. (2000). The use of phytoplankton pigments for identifying and quantifying phytoplankton groups in coastal areas: testing the influence of light and nutrients on pigment/chlorophyll a ratios. Marine Ecology progress series, 192, 49-63.

[13] Morel, A., & Prieur, L. (1977). Analysis of variations in ocean color 1. Limnology and oceanography, 22(4), 709-722.

[14] Bidigare, R. R., Ondrusek, M. E., Morrow, J. H., & Kiefer, D. A. (1990, September). In-vivo absorption properties of algal pigments. In Ocean Optics X (Vol. 1302, pp. 290-302). International Society for Optics and Photonics.

[15] Bricaud, A., Claustre, H., Ras, J., & Oubelkheir, K. (2004). Natural variability of phytoplanktonic absorption in oceanic waters: Influence of the size structure of algal

populations. Journal of Geophysical Research: Oceans, 109(C11).

[16] Gordon, H. R., Brown, O. B., Evans, R. H., Brown, J. W., Smith, R. C., Baker, K. S., & Clark, D. K. (1988). A semianalytic radiance model of ocean color. Journal of Geophysical Research: Atmospheres, 93(D9), 10909-10924.

[17] Gordon, H. R., Clark, D. K., Brown, J. W., Brown, O. B., Evans, R. H., & Broenkow, W. W. (1983). Phytoplankton pigment concentrations in the Middle Atlantic Bight: comparison of ship determinations and CZCS estimates. Applied optics, 22(1), 20-36.

[18] Morel, A., & Antoine, D. (2000). Pigment index retrieval in Case 1 waters. MERIS ATBD, 2.

[19] O'Reilly, J. E., Maritorena, S., Mitchell, B. G., Siegel, D. A., Carder, K. L., Garver, S. A., ... & McClain, C. (1998). Ocean color chlorophyll algorithms for SeaWiFS. Journal of Geophysical Research: Oceans, 103(C11), 24937-24953.

[20] O'Reilly, J. E., Maritorena, S., Siegel, D. A., O'Brien, M. C., Toole, D., Mitchell, B. G., ... & Hooker, S. B. (2000). Ocean color chlorophyll a algorithms for SeaWiFS, OC2, and OC4: Version 4. SeaWiFS postlaunch calibration and validation analyses, Part, 3, 9-23.

[21] Sosik, H. M., & Mitchell, B. G. (1995). Light absorption by phytoplankton, photosynthetic pigments and detritus in the California Current System. Deep Sea Research Part I: Oceanographic Research Papers, 42(10), 1717-1748.

[22] Laws, E. A., & Bannister, T. T. (1980). Nutrient-and light-limited growth of Thalassiosira fluviatilis in continuous culture, with implications for phytoplankton growth in the ocean 1. Limnology and Oceanography, 25(3), 457-473.

[23] Lubart, R., Lavi, R., Friedmann, H., & Rochkind, S. (2006). Photochemistry and photobiology of light absorption by living cells. Photomedicine and Laser Therapy, 24(2), 179-185.

[24] Mueller, J. L. (2002). Ocean optics protocols for satellite ocean color sensor validation.

[25] Kirkpatrick, G. J., Millie, D. F., Moline, M. A., & Schofield, O. (2000). Optical discrimination of a phytoplankton species in natural mixed populations. Limnology and Oceanography, 45(2), 467-471.

[26] Subramaniam, A., Brown, C. W., Hood, R. R., Carpenter, E. J., & Capone, D. G. (2001). Detecting Trichodesmium blooms in SeaWiFS imagery. Deep Sea Research Part II: Topical Studies in Oceanography, 49(1-3), 107-121.

[27] Tomlinson, M. C., Wynne, T. T., & Stumpf, R. P. (2009). An evaluation of remote sensing techniques for enhanced detection of the toxic dinoflagellate, Karenia brevis. Remote Sensing of Environment, 113(3), 598-609.

[28] Bricaud, A., & Stramski, D. (1990). Spectral absorption coefficients of living phytoplankton and nonalgal biogenous matter: A comparison between the Peru upwelling areaand the Sargasso Sea. Limnology and oceanography, 35(3), 562-582.

[29] Hoepffner, N., & Sathyendranath, S. (1991). Effect of pigment composition on absorption properties of phytoplankton. Mar. Ecol. Prog. Ser, 73(1), 1-23.

[30] Moisan, J. R., Moisan, T. A., & Linkswiler, M. A. (2011). An inverse modeling approach to estimating phytoplankton pigment concentrations from phytoplankton absorption spectra.

Journal of Geophysical Research: Oceans, 116(C9).

[31] Hoepffner, N., & Sathyendranath, S. (1993). Determination of the major groups of phytoplankton pigments from the absorption spectra of total particulate matter. Journal of Geophysical Research: Oceans, 98(C12), 22789-22803.

[32] Lohrenz, S. E., Weidemann, A. D., & Tuel, M. (2003). Phytoplankton spectral absorption as influenced by community size structure and pigment composition. Journal of Plankton Research, 25(1), 35-61.

[33] Chase, A., Boss, E., Zaneveld, R., Bricaud, A., Claustre, H., Ras, J., ... & Westberry, T. K. (2013). Decomposition of in situ particulate absorption spectra. Methods in Oceanography, 7, 110-124.

[34] Wang, G., Lee, Z., Mishra, D. R., & Ma, R. (2016). Retrieving absorption coefficients of multiple phytoplankton pigments from hyperspectral remote sensing reflectance measured over cyanobacteria bloom waters. Limnology and Oceanography: Methods, 14(7), 432-447.

[35] Wang, G., Lee, Z., & Mouw, C. (2017). Multi-spectral remote sensing of phytoplankton pigment absorption properties in cyanobacteria bloom waters: a regional example in the western basin of Lake Erie. Remote Sensing. 9, no. 12: 1309. doi:10.3390/rs9121309.

[36] Babin, M. (2008). Phytoplankton fluorescence: theory, current literature and in situ measurement. Real-time coastal observing systems for marine ecosystem dynamics and harmful algal blooms, edited by: Babin, M., Roesler, C., and Cullen, JJ, 237-280.

[37] Neville, R. A., & Gower, J. F. R. (1977). Passive remote sensing of phytoplankton via chlorophyll α fluorescence. Journal of Geophysical Research, 82(24), 3487-3493.

[38] Huot, Y., Brown, C. A., & Cullen, J. J. (2005). New algorithms for MODIS sun-induced chlorophyll fluorescence and a comparison with present data products. Limnology and Oceanography: Methods, 3(2), 108-130.

[39] Prézelin, B. B., & Alberte, R. S. (1978). Photosynthetic characteristics and organization of chlorophyll in marine dinoflagellates. Proceedings of the National Academy of Sciences, 75(4), 1801-1804.

[40] Falkowski, P.G. and Owens T.G. (1980). Light-shade adaptation - 2 strategies in marine-phytoplankton. Plant Physiol. 66(4): 592-595.

[41] Cullen, J. J. (2008). Observation and prediction of harmful algal blooms. Real-time coastal observing system for marine ecosystem dynamics and harmful algal blooms: theory, Instrument and Modeling. UNESCO, Paris.

[42] Du, H., R.-C. A. Fuh, J. Li, L. A. Corkan and J. S. Lindsey (1998) PhotochemCAD: A computer-aided design and research tool in photochemistry. Photochem. Photobiol. 68, 141-142.

[43] Dixon, J. M., M. Taniguchi and J. S. Lindsey (2005), "PhotochemCAD 2. A Refined Program with Accompanying Spectral Databases for Photochemical Calculations, Photochem. Photobiol., 81, 212-213.

[44] Trees, C. C., Kennicutt II, M. C., & Brooks, J. M. (1985). Errors associated with the standard fluorimetric determination of chlorophylls and phaeopigments. Marine Chemistry, 17(1), 1-12.

[45] Trees, C. C., Clark, D. K., Bidigare, R. R., Ondrusek, M. E., & Mueller, J.

L. (2000). Accessory pigments versus chlorophyll a concentrations within the euphotic zone: A ubiquitous relationship. Limnology and Oceanography, 45(5), 1130-1143.

[46] Kumari, B. (2005). Comparison of high performance liquid chromatography and fluorometric ocean colour pigments. Journal of the Indian Society of Remote Sensing, 33(4), 541-546.

[47] Lorenzen, C. J., & Jeffrey, S. W. (1980). Determination of chlorophyll in seawater. Unesco tech. pap. mar. sci, 35(1), 1-20.

[48] Abaychi, J. K., & Riley, J. P. (1979). The determination of phytoplankton pigments by high-performance liquid chromatography. Analytica Chimica Acta, 107, 1-11.

[49] Mantoura, R. F. C., & Llewellyn, C. A. (1983). The rapid determination of algal chlorophyll and carotenoid pigments and their breakdown products in natural waters by reverse-phase high-performance liquid chromatography. Analytica Chimica Acta, 151, 297-314.

[50] Aneeshkumar, N., & Sujatha, C. H. (2012). Biomarker pigment signatures in Cochin back water system–A tropical estuary south west coast of India. Estuarine, Coastal and Shelf Science, 99, 182-190.

[51] Barlow, R., Kyewalyanga, M., Sessions, H., Van den Berg, M., & Morris, T. (2008). Phytoplankton pigments, functional types, and absorption properties in the Delagoa and Natal Bights of the Agulhas ecosystem. Estuarine, Coastal and Shelf Science, 80(2), 201-211.

[52] Paerl, H. W., Valdes, L. M., Pinckney, J. L., Piehler, M. F., Dyble, J., & Moisander, P. H. (2003). Phytoplankton photopigments as indicators of estuarine and coastal eutrophication. BioScience, 53(10), 953-964.

[53] Sieburth, J. M., Smetacek, V., & Lenz, J. (1978). Pelagic ecosystem structure: Heterotrophic compartments of the plankton and their relationship to plankton size fractions 1. Limnology and oceanography, 23(6), 1256-1263.

[54] Vidussi, F., Claustre, H., Manca, B. B., Luchetta, A., & Marty, J. C. (2001). Phytoplankton pigment distribution in relation to upper thermocline circulation in the eastern Mediterranean Sea during winter. Journal of Geophysical Research: Oceans, 106(C9), 19939-19956.

[55] Hovis, W. A., Clark, D. K., Anderson, F., Austin, R. W., Wilson, W. H., Baker, E. T., ... & Sturm, B. (1980). Nimbus-7 Coastal Zone Color Scanner: system description and initial imagery. Science, 210(4465), 60-63.

[56] Gordon, H. R., Clark, D. K., Mueller, J. L., & Hovis, W. A. (1980). Phytoplankton pigments from the Nimbus-7 Coastal Zone Color Scanner: comparisons with surface measurements. Science, 210(4465), 63-66.

[57] Smith, R. C., & Baker, K. S. (1982). Oceanic chlorophyll concentrations as determined by satellite (Nimbus-7 Coastal Zone Color Scanner). Marine Biology, 66(3), 269-279.

[58] Mitchell, B. G. (1994). Coastal zone color scanner retrospective. Journal of Geophysical Research: Oceans, 99(C4), 7291-7292.

[59] Simis, S. G., Peters, S. W., & Gons, H. J. (2005). Remote sensing of the cyanobacterial pigment phycocyanin in turbid inland water. Limnology and Oceanography, 50(1), 237-245.

[60] Mishra, S., Mishra, D. R., Lee, Z., & Tucker, C. S. (2013). Quantifying

cyanobacterial phycocyanin concentration in turbid productive waters: A quasi-analytical approach. Remote Sensing of Environment, 133, 141-151.

[61] Dierssen, H. M. (2010). Perspectives on empirical approaches for ocean color remote sensing of chlorophyll in a changing climate. Proceedings of the National Academy of Sciences, 107(40), 17073-17078.

[62] Hu, C., & Campbell, J. (2014). Oceanic chlorophyll-a content. In Biophysical applications of satellite remote sensing (pp. 171-203). Springer, Berlin, Heidelberg.

[63] Hu, C., Lee, Z., & Franz, B. (2012). Chlorophyll aalgorithms for oligotrophic oceans: A novel approach based on three-band reflectance difference. Journal of Geophysical Research: Oceans, 117(C1).

[64] Kahru, M., & Mitchell, B. G. (1999). Empirical chlorophyll algorithm and preliminary SeaWiFS validation for the California Current. International Journal of Remote Sensing, 20(17), 3423-3429.

[65] Hu, C., Feng, L., & Lee, Z. (2013). Uncertainties of SeaWiFS and MODIS remote sensing reflectance: Implications from clear water measurements. Remote Sensing of Environment, 133, 168-182.

[66] Szeto, M., Werdell, P. J., Moore, T. S., & Campbell, J. W. (2011). Are the world's oceans optically different?. Journal of Geophysical Research: Oceans, 116(C7).

[67] Pan, X., Mannino, A., Russ, M. E., Hooker, S. B., & Harding Jr, L. W. (2010). Remote sensing of phytoplankton pigment distribution in the United States northeast coast. Remote Sensing of Environment, 114(11), 2403-2416.

[68] Pan, X., Mannino, A., Marshall, H. G., Filippino, K. C., & Mulholland, M. R. (2011). Remote sensing of phytoplankton community composition along the northeast coast of the United States. Remote Sensing of Environment, 115(12), 3731-3747.

[69] Carder, K. L., Chen, F. R., Lee, Z. P., Hawes, S. K., & Kamykowski, D. (1999). Semianalytic Moderate-Resolution Imaging Spectrometer algorithms for chlorophyll a and absorption with bio-optical domains based on nitrate-depletion temperatures. Journal of Geophysical Research: Oceans, 104(C3), 5403-5421.

[70] IOCCG (2006). Remote Sensing of Inherent Optical Properties: Fundamentals, Tests of Algorithms, and Applications. Lee, Z.-P. (ed.), Reports of the International Ocean-Colour Coordinating Group, No. 5, IOCCG, Dartmouth, Canada

[71] Lee, Z., Carder, K. L., & Arnone, R. A. (2002). Deriving inherent optical properties from water color: a multiband quasi-analytical algorithm for optically deep waters. Applied optics, 41(27), 5755-5772.

[72] Maritorena, S., Siegel, D. A., & Peterson, A. R. (2002). Optimization of a semianalytical ocean color model for global-scale applications. Applied optics, 41(15), 2705-2714.

[73] Sathyendranath, S., Prieur, L., & Morel, A. (1989). A three-component model of ocean colour and its application to remote sensing of phytoplankton pigments in coastal waters. International Journal of Remote Sensing, 10(8), 1373-1394.

[74] Wang, G., Lee, Z., & Mouw, C. B. (2018). Concentrations of multiple phytoplankton pigments in the global oceans obtained from satellite ocean color measurements with MERIS. Applied Sciences, 8(12), 2678.

[75] Wozniak, B., Dera, J., Ficek, D., Majchrowski, R., Kaczmarek, S., Ostrowska, M., & Koblentz-Mishke, O. I. (2000). Model of the in vivo spectral absorption of algal pigments. Part 1. Mathematical apparatus. Oceanologia, 42(2).

[76] Moisan, T. A., Rufty, K. M., Moisan, J. R., & Linkswiler, M. A. (2017). Satellite observations of phytoplankton functional type spatial distributions, phenology, diversity, and ecotones. Frontiers in Marine Science, 4, 189.

[77] Sathyendranath, S., Cota, G., Stuart, V., Maass, H., & Platt, T. (2001). Remote sensing of phytoplankton pigments: a comparison of empirical and theoretical approaches. International Journal of Remote Sensing, 22(2-3), 249-273.

[78] Gitelson, A. A., Schalles, J. F., & Hladik, C. M. (2007). Remote chlorophyll-a retrieval in turbid, productive estuaries: Chesapeake Bay case study. Remote Sensing of Environment, 109(4), 464-472.

[79] Abbott, M. R., & Letelier, R. M. (1999). Algorithm theoretical basis document chlorophyll fluorescence (MODIS product number 20). NASA (http://www. modis. gsfc. nasa. gov/ data/atbd).

[80] Garver, S. A., & Siegel, D. A. (1997). Inherent optical property inversion of ocean color spectra and its biogeochemical interpretation: 1. Time series from the Sargasso Sea. Journal of Geophysical Research: Oceans, 102(C8), 18607-18625.

[81] Siegel, D. A., Maritorena, S., Nelson, N. B., Hansell, D. A., & Lorenzi-Kayser, M. (2002). Global distribution and dynamics of colored dissolved and detrital organic materials. Journal of Geophysical Research: Oceans, 107(C12), 21-1.

[82] Siegel, D. A., Maritorena, S., Nelson, N. B., & Behrenfeld, M. J. (2005). Independence and interdependencies among global ocean color properties: Reassessing the bio-optical assumption. Journal of Geophysical Research: Oceans, 110(C7).

[83] Werdell, P. J., Franz, B. A., Bailey, S. W., Feldman, G. C., Boss, E., Brando, V. E., ... & Loisel, H. (2013). Generalized ocean color inversion model for retrieving marine inherent optical properties. Applied optics, 52(10), 2019-2037.

[84] Uitz, J., Claustre, H., Morel, A., & Hooker, S. B. (2006). Vertical distribution of phytoplankton communities in open ocean: An assessment based on surface chlorophyll. Journal of Geophysical Research: Oceans, 111(C8).

[85] Hirata, T., Aiken, J., Hardman-Mountford, N., Smyth, T. J., & Barlow, R. G. (2008). An absorption model to determine phytoplankton size classes from satellite ocean colour. Remote Sensing of Environment, 112(6), 3153-3159.

[86] Brewin, R. J., Sathyendranath, S., Hirata, T., Lavender, S. J., Barciela, R. M., & Hardman-Mountford, N. J. (2010). A three-component model of phytoplankton size class for the Atlantic Ocean. Ecological Modelling, 221(11), 1472-1483.

[87] Kostadinov, T. S., Siegel, D. A., & Maritorena, S. (2009). Retrieval of the particle size distribution from satellite ocean color observations. Journal of Geophysical Research: Oceans, 114(C9).

[88] Ciotti, A. M., & Bricaud, A. (2006). Retrievals of a size parameter for phytoplankton and spectral light absorption by colored detrital matter from water-leaving radiances at SeaWiFS channels in a continental

shelf region off Brazil. Limnology and Oceanography: Methods, 4(7), 237-253.

[89] Mouw, C. B., & Yoder, J. A. (2010). Optical determination of phytoplankton size composition from global SeaWiFS imagery. Journal of Geophysical Research: Oceans, 115(C12).

[90] Devred, E., Sathyendranath, S., Stuart, V., & Platt, T. (2011). A three component classification of phytoplankton absorption spectra: Application to ocean-color data. Remote Sensing of Environment, 115(9), 2255-2266.

[91] Bricaud, A., Ciotti, A. M., & Gentili, B. (2012). Spatial-temporal variations in phytoplankton size and colored detrital matter absorption at global and regional scales, as derived from twelve years of SeaWiFS data (1998-2009). Global Biogeochemical Cycles, 26(1).

[92] Alvain, S., Moulin, C., Dandonneau, Y., & Bréon, F. M. (2005). Remote sensing of phytoplankton groups in case 1 waters from global SeaWiFS imagery. Deep Sea Research Part I: Oceanographic Research Papers, 52(11), 1989-2004.

[93] Mustapha, Z. B., Alvain, S., Jamet, C., Loisel, H., & Dessailly, D. (2014). Automatic classification of water-leaving radiance anomalies from global SeaWiFS imagery: application to the detection of phytoplankton groups in open ocean waters. Remote sensing of environment, 146, 97-112.

[94] Bracher, A., Vountas, M., Dinter, T., Burrows, J. P., Röttgers, R., & Peeken, I. (2009). Quantitative observation of cyanobacteria and diatoms from space using PhytoDOAS on SCIAMACHY data. Biogeosciences, 6, 751-764.

[95] Sadeghi, A., Dinter, T., Vountas, M., Taylor, B., Peeken, I., Altenburg Soppa, M., & Bracher, A. (2012). Improvements to the PhytoDOAS method for identification of coccolithophores using hyper-spectral satellite data. Ocean Science, 8(6), 1055-1070.

[96] Raitsos, D. E., Lavender, S. J., Maravelias, C. D., Haralabous, J., Richardson, A. J., & Reid, P. C. (2008). Identifying four phytoplankton functional types from space: An ecological approach. Limnology and oceanography, 53(2), 605-613.

[97] Palacz, A., John, M. S., Brevin, R. J. W., Hirata, T., & Gregg, W. W. (2013). Distribution of phytoplankton functional types in high-nitrate low-chlorophyll waters in a new diagnostic ecological indicator model. Biogeosciences, 10(11), 7553-7574.

[98] Kostadinov, T. S., Cabré, A., Vedantham, H., Marinov, I., Bracher, A., Brewin, R. J., ... & Mouw, C. (2017). Inter-comparison of phytoplankton functional type phenology metrics derived from ocean color algorithms and earth system models. Remote sensing of environment, 190, 162-177.

[99] IOCCG (2014). Phytoplankton Functional Types from Space. Sathyendranath, S. (ed.), Reports of the International Ocean-Colour Coordinating Group, No. 15, IOCCG, Dartmouth, Canada.

[100] Wright, S. W., & Jeffrey, S. W. (2006). Pigment markers for phytoplankton production. In Marine organic matter: biomarkers, isotopes and DNA (pp. 71-104). Springer, Berlin, Heidelberg.

[101] Roy, S., Alam, S., & Chattopadhyay, J. (2006). Competing effects of toxin-producing phytoplankton on overall plankton populations in the Bay of Bengal. Bulletin of Mathematical Biology, 68(8), 2303-2320.

[102] Barlow, R. G., Aiken, J., Holligan, P. M., Cummings, D. G., Maritorena, S., & Hooker, S. (2002). Phytoplankton pigment and absorption characteristics along meridional transects in the Atlantic Ocean. Deep Sea Research Part I: Oceanographic Research Papers, 49(4), 637-660.

[103] Gibb, S. W., Barlow, R. G., Cummings, D. G., Rees, N. W., Trees, C. C., Holligan, P., & Suggett, D. (2000). Surface phytoplankton pigment distributions in the Atlantic Ocean: an assessment of basin scale variability between 50 N and 50 S. Progress in Oceanography, 45(3-4), 339-368.

[104] Moreno, D. V., Marrero, J. P., Morales, J., García, C. L., Úbeda, M. V., Rueda, M. J., & Llinás, O. (2012). Phytoplankton functional community structure in Argentinian continental shelf determined by HPLC pigment signatures. Estuarine, Coastal and Shelf Science, 100, 72-81.

[105] Vijayan, A. K., Yoshikawa, T., Watanabe, S., Sasaki, H., Matsumoto, K., Saito, S. I., ... & Furuya, K. (2009). Influence of non-photosynthetic pigments on light absorption and quantum yield of photosynthesis in the western equatorial Pacific and the subarctic North Pacific. Journal of oceanography, 65(2), 245-258.

[106] IOCCG (2019). Uncertainties in Ocean Colour Remote Sensing. Mélin F. (ed.), IOCCG Report Series, No. 18, International Ocean Colour Coordinating Group, Dartmouth, Canada. http://dx.doi.org/10.25607/OBP696

[107] Werdell, P. J., Behrenfeld, M. J., Bontempi, P. S., Boss, E., Cairns, B., Davis, G. T., & Knobelspiesse, K. D. (2019). The Plankton, Aerosol, Cloud, ocean Ecosystem mission: status, science, advances. Bulletin of the American Meteorological Society, 100(9), 1775-1794.

[108] Behrenfeld, M. J., Hu, Y., O'Malley, R. T., Boss, E. S., Hostetler, C. A., Siegel, D. A., ... & Rodier, S. (2017). Annual boom–bust cycles of polar phytoplankton biomass revealed by space-based lidar. Nature Geoscience, 10(2), 118-122.

[109] Hooker, S. B., & McClain, C. R. (2000). The calibration and validation of SeaWiFS data. Progress in Oceanography, 45(3-4), 427-465.

[110] Nair, A., Sathyendranath, S., Platt, T., Morales, J., Stuart, V., Forget, M. H., ... & Bouman, H. (2008). Remote sensing of phytoplankton functional types. Remote Sensing of Environment, 112(8), 3366-3375.

[111] Hallegraeff, G. M. (2010). Ocean climate change, phytoplankton community responses, and harmful algal blooms: a formidable predictive challenge 1. Journal of phycology, 46(2), 220-235.

[112] Hostetler, C. A., Behrenfeld, M. J., Hu, Y., Hair, J. W., & Schulien, J. A. (2018). Spaceborne lidar in the study of marine systems. Annual review of marine science, 10, 121-147.

[113] Churnside, J. H., & Shaw, J. A. (2020). Lidar remote sensing of the aquatic environment. Applied Optics, 59(10), C92-C99.

[114] Moore, T. S., Churnside, J. H., Sullivan, J. M., Twardowski, M. S., Nayak, A. R., McFarland, M. N., ... & Ruberg, S. A. (2019). Vertical distributions of blooming cyanobacteria populations in a freshwater lake from LIDAR observations. Remote Sensing of Environment, 225, 347-367.

Chapter 4

First Report on the Diversity of Epizoic Algae in Larval of Shellfish Gastropod *Aliger gigas*

Rocío Elizabeth Muciño-Márquez, Dalila Aldana-Aranda, María Guadalupe Figueroa-Torres and Oscar Hernández-Almeida

Abstract

Epibiosis occur frequently on the shells of some marine crustaceans, which often serve as substrate for various species of algae, there is few information on the associations between these. The objective of this study was to determine if the gastropod mollusk *Aliger gigas* (formerly *Lobatus gigas*) in larval had some sort of the association with algal. To the above was carried out collecting egg masses in the environment, the larvae were cultivated in seawater filtered 5 μm. The algal material found was observed in electron microscopy, for its identification and quantification. We analyzed 60 larvae aged 2–44 days for analyzing the structure of the shell and its epibionts. Of the larvae analyzed, 50 larvae presented epizoic. The algae community consisted of 28 taxa, and composed of 25 diatoms (Bacillariophyta) and three cyanophytes (Cyanobacteria). The H′ diversity values fluctuated between 0.2 a 1.2. The dominant and frequent species were formed by diatoms: *Nitzschia panduriformis var. minor*, *Halamphora* sp. and *Cyclophora* sp.

Keywords: cyanophyte, diatom, epibiont, *Aliger gigas*

1. Introduction

Epibiotic is a type of association in which an organism lives on the surface layer of another organism called basibiont, these nonparasitic organisms are known as epibionts [1, 2]. The shells of gastropod and bivalve mollusks represent a suitable habitat for the settlement of various species of algae, viruses or fungi [3–6]. Different studies have focused on epiphytic diatoms of grasses and marine macroalgae [7–9]; in copepods of the species *Farranula gibbula*, the epibiotic diatom *Pseudohimantidium pacificum* has been observed [10]. Very little information exists on symbiotic associations between algae and crustaceans or marine planktonic mollusks, being able to cite what was observed in *Peringia ulvae* (formerly *Hydrobia ulvae*) and diatoms *Cocconeis placentula* and *Achnanthes lemmermannii*, also cyanophytes and bacteria in its Shell [5]. Based on the above, the objective of this work was to identify epizoic species present in the shells of the larval stages of the marine gastropod mollusk, *Aliger gigas*.

2. Material and methods

An egg mass of *Aliger gigas* was incubated in filtered seawater with a 5 μm mesh, until hatching. Later, the larvae were cultured with seawater filtered with a 50 μm mesh and fed with *Nannochloropsis oculata* (Ochrophyta, Eustigmatophyceae) at a concentration of 1 000 cells per larva, at a density of 100 larvae/L. The larvae were fixed in glutaraldehyde, cacodylate and dehydrated in alcohols from 70 to 100% and dried at a critical point. The shells of 60 larvae between 2 and 44 days old were processed. The specimens were observed in a JEOL field emission scanning electron microscope (JSM-7600F), of the National Laboratory of Nano and Biomaterials of Cinvestav IPN Mérida, the presence of epizoic algae was analyzed and to its quantification was carried out. For the identification of phytoplankton, the works of [11–16], among others. The AlgaeBase system was consulted to verify accepted taxonomic names [17].

To obtain the relative abundance index, the proportion of abundance of each species (organism number) was quantified in relation to the total abundance of organisms counted in each larva of different ages [18]. The contribution of the abundance of the epizoic algae species of each larva was determined by means of the SIMPER analysis [19]. This analysis determines the species that most contribute to the similarity between sample. A cumulative similarity discrimination value of 90% was applied. Based on the composition and abundance of the epizoic algae species, the community was characterized by the following descriptors: to evaluate the diversity, the species richness of Margalef (S), the Shannon-Wiener index (H′) and Pielou's equity (J') considering to according to in accordance with [20], through the ODI program of the Interdisciplinary Center for Marine Sciences, Department of Plankton.

To obtain dominance of the species, an Olmstead & Tukey test was used [21]. The dominant, constant, occasional and rare species were determined from the relationship between the densities of the organisms and their frequencies of appearance. The statistical programs used were Primer-E and R.

3. Results

Of the 60 specimens of *A. gigas* larvae analyzed, 83% presented epizoic algae. The epizoic algae community consisted of 28 taxa, made up of 25 diatoms and three cyanophytes. It should be noted that one of the recorded diatom species *Cylindrotheca closterium* is considered a species that can be harmful and forms algal blooms (**Table 1**).

3.1 Specific diversity

The diversity values H′ fluctuated between 0.9 and 1.2. The 28-day-old pre-metamorphic larval shells presented the highest value of H′ 1.2 with an equity of J′ 0.4, and a species richness of S 14. These values of H′ 1.2 with a J 0.5 and an S 9, were slightly higher in the 30-day-old larvae, which already had foot formation. For the 20-day-old larvae, H 'was 1.1, J' was 0.4 and S was 11 and in the 18-day-old larvae, H 'was 0.9, J' was 0.4 and S was 9 species (**Figure 1**, **Table 1**). Following the same behavior, the youngest veliger larvae, 8 days old, presented the lowest diversity with values of H 'of 0.2, J' of 0.42 and S of 4 species (**Figure 1**, **Table 1**).

3.2 Dominant species

Based on the Olmstead and Tukey test, the epizoic algae community consisted of 17 rare species, followed by five common, three abundant and three dominant

Larva age in days	**8**	**15**	**16**	**18**	**20**	**26**	**28**	**30**	**36**	**42**	**%FR**	**At.**
Number of larvae analyzed	**4**	**2**	**9**	**1**	**2**	**2**	**10**	**2**	**6**	**6**		
Bacillariophyta												
Amphora sp.	0	0	0	0	0	0	0	2.5	0	0	10	R
Cocconeis lineata	0	0	0	0	0.1	1.5	0	3	0	0	30	R
Cocconeis scutellum	0	0	0	0	0.1	2	0.1	0	0	0	30	O
Craspedostauros sp.	0	5	0	0	0	0	12	0	0.1	0	30	R
Cyclophora sp.	0	66	10	3	5	1.9	0.3	15	76	70	90	D
Cylindrotheca closterium	1	7.5	1.5	5	4.4	1.3	0.3	0	0	0	70	C
Entomoneis paludosa	0	0	19	3.6	7	17	16	11	8.5	4	80	C
Halamphora coffeaeformis	0	0	0	0	0	0	0	0	0.1	0.1	20	R
Halamphora sp.	0	1.7	47	30	10	7	0.2	0	0	0	70	D
Haslea tsukamotoi	0	0	0	0	0	1	0.1	7.5	0	1	40	R
Hippodonta pseudacceptata	0	0	0	0	54.3	0	0	0	0	0	10	R
Hyalosynedra sp.	0	0	0	0	0	0.4	0.1	0	0	1.2	30	R
Licmophora sp.	0	0	0	0.3	0	0	0	0	0	0	10	R
Navicula radiosa	0	0	0	2.6	0	19	1	9	0	0	40	O
Nitzschia dissipata	0	0	0	0	0.1	0	0.1	0	0	0	20	R
Nitzschia inconspicua	0	0	0	0	0	8	0	0	0	2.8	20	R
Nitzschia linearis	0	0	0	0	0	0	0	0	0	0.3	10	R
Nitzschia microcephala	0	0	0.4	0	0	0	0.1	0	0	0	20	R
Nitzschia panduriformis var. minor	64	18	8	51	9	4.5	1.1	20	7.5	0.2	100	D
Nitzschia sp.	0	0	3.6	4	5	0	0	0	0	0	30	O
Pleurosigma sp.	5	0	0	0.5	0	0	0	0	0	0	20	R

Psammodictyon panduriforme	0	0	2.2	0	0	0	0	0	0	0	10	R
Pseudachnanthidium sp.	0	0	0	0	0	0.3	0	0	0	0	10	R
Scalariella sp.	0	0	8	0	0	0	0	0	0.2	0.1	30	R
Stephanodiscus minutulus	0	0	0.3	0	0	0	0	0	0	0	10	R
Cyanophyta												
Haloleptolyngbya sp.	0	0	0	0	5	32	68	30	7.6	20	60	C
Richelia intracellularis	0	0	0	0	0	0	0	2	0	0	10	R
Arthrospira sp.	0	1.8	0	0	0	4.1	0.6	0	0	0.3	40	R
S	**4**	**6**	**10**	**9**	**11**	**14**	**14**	**9**	**7**	**11**		
H′	**0.2**	**0.7**	**0.4**	**0.9**	**1.1**	**0.7**	**1.2**	**1.2**	**0.5**	**0.7**		
J′	**0.2**	**0.4**	**0.2**	**0.4**	**0.4**	**0.3**	**0.4**	**0.5**	**0.3**	**0.3**		

Table 1.
Percentage of the relative abundance of the epizoic algae community in the shells of Aliger gigas larvae. % FR: Percentage of the relative frequency. At: Attribute, D: D: Dominant, C: Constant, O: Occasional and R: Rare. S: Species richness, H′: Diversity and J′: Equity.

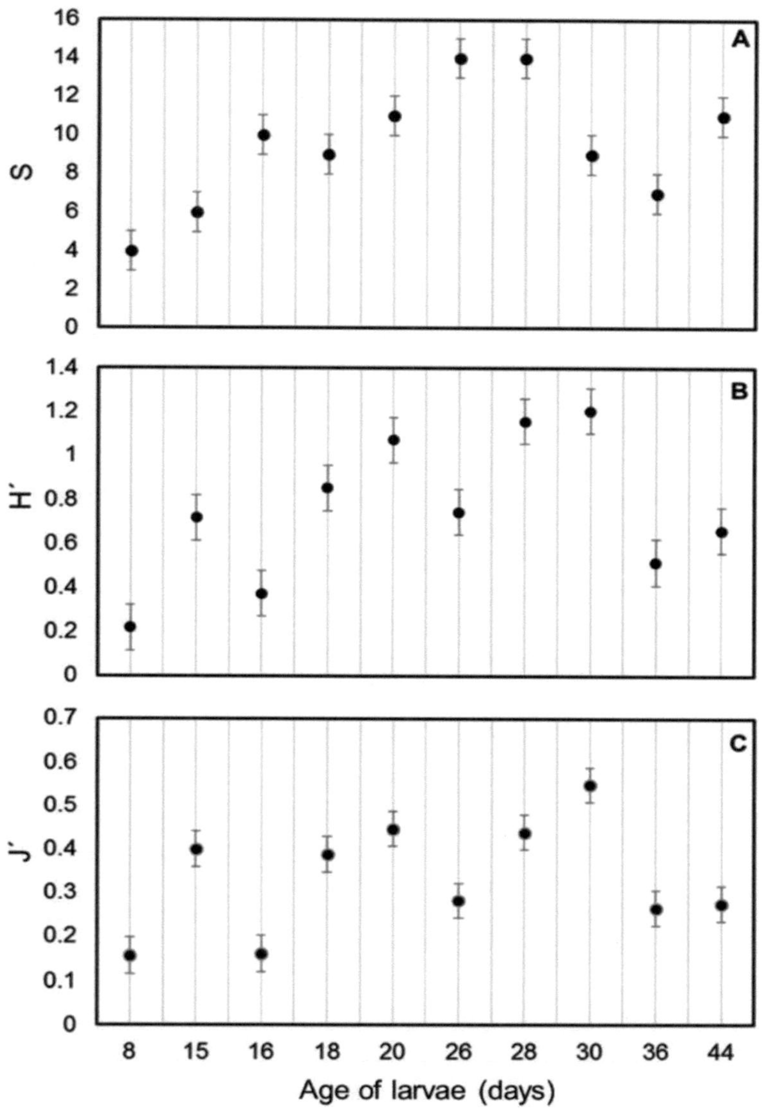

Figure 1.
(A) Species richness (S), (B) diversity (H') and (C) equity (J') of the epizoic algae community in the shells of Aliger gigas larvae.

species (**Table 1**). The dominant species were made up by the diatom *Nitzschia panduriformis var. minor*, whose highest relative abundance was 68% in 8-day-old larvae; *Hippodonta pseudacceptata* with a relative abundance of 54% in 20-day-old larvae; *Halamphora* sp. with a relative abundance of 47% in 16-day-old larvae and *Cyclophora* sp. with 70 and 76% in larvae of 36 and 44 days respectively (**Figure 2**, **Table 1**). In addition, of the cyanophyte *Haloleptolyngbya* sp. with a relative abundance of 68% in 28-day-old larvae (**Figure 2, Table 1**).

3.3 Characteristics of *Aliger gigas* larvae, observed in the development of this work

The two to five-day old larvae have a shell formed by two turns in a spiral presenting small granule at the apex and a velum characterized by having two lobes and the right tentacle, corresponding to a young veliger larva. The shell of eight-day-old

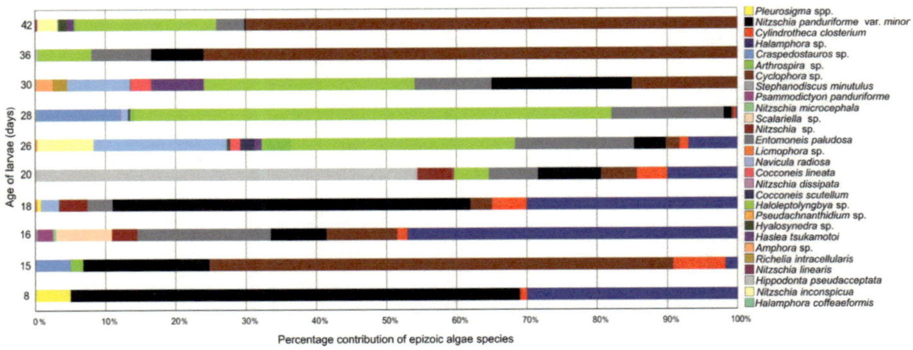

Figure 2.
SIMPER analysis, percentage contribution of epizoic algae species from the shells of the larvae of the marine gastropod mollusk Aliger gigas.

larvae is characterized by having three coils, showing well-defined lines of ornamentation on the body of the shell. As regards its development; the velum has four lobes; with the right tentacle well differentiated and the formation of the left tentacle.

The 15–18-day old larvae have a carapace with three and a half turns in the spiral, the body has four parallel lines ending at the end of the siphon channel and is highly ornamented. The velum has six lobes, tentacles, and proboscis. In larvae from 20 to 28 day of development, their shell has three and a half turns, with a band of uniform striations on the body of the shell, the radula is already observed and the velum begins the process of reabsorption therefore the larva begins to have a creeping shape beginning its benthic phase. It is a stage known as a precompetent larva. Regarding larvae between 30 and 44 days old, at this stage the larvae are ready to metamorphosis (30 days). Post-metamorphic larvae or post larvae of 44 days, present a foot with an active crawling behavior. The shell is characterized by presenting four turns, with a well-developed band of striae, the proboscide and the radula are present and active, as is the foot with its operculum.

4. Discussion

Diatoms have been reported in the literature as the main group of epizoic microalgae species attached to different types of animals that can be copepods [10, 22–24]; cladocerans [25], hydrozoans [26–27], krill [28] even in whales [29–30]. Diatoms are also present in diving birds [31–32] and reptiles such as crocodiles [33].

As mentioned by [34], the first phase of the colonization of a substrate occurs mainly by bacteria with diatoms, fungi and protozoa; which generate a film on the surface of the basibiont. The results of this study showed that the shells of larvae of the marine gastropod mollusk, *Aliger gigas*, provide an adequate and frequent substrate for the settlement of epizoic microalgae, in the case of diatoms and cyanophytes. A dominance of diatom species was also observed. *Nitzschia panduriformis var. minor* was reported in 8-day-old larvae, *Halamphora* sp. was present in 16-day-old larvae, *Hippodonta pseudacceptata*, in larvae of 20 days and *Cyclophora* sp., in larvae of 36 and 44 days. Obtaining low values of diversity H'(0.2 to 0.9) and J' (0.2 to 0.5), in these phases of the larvae. Margalef [35] mentions that diversity is low when there is a dominance of some species.

Likewise, it was found that the diatom and cyanophyte populations were not stable. It is interesting to note that, although the larvae were under the same culture conditions, in the larvae of different ages, the structure of their epizoic microalgae community changed. In larvae less than 20 days old, the cells of the dominant microalgae were shed from the larvae shells allowing the colonization of other microalgae; on the other hand, in larvae of 20 to 30 days a higher H′ diversity of 1.0 to 1.2 was reported. In another study with gastropods, carried out by [6] reviewed the shells of seven gastropods (*Alvania lineata*, *Bittium reticulatum*, *Clanculus cruciatus*, *Columbella rustica*, *Gibbula adansoni*, *Nassarius incrassatus* and *Jujubinus striatus*), reporting a richness of 19 to 25 species and a high J' equity of 0.70 to 0.80 and [36] in *Lepidochelys olivacea* (olive ridley) shells, recorded a diversity of 1.1 to 2.1 and a high J' equity of 0.56 to 0.86.

From what was observed in this study, the size and structure of the shell of *A. gigas* larvae on the different days of development provide a substrate for the epizoic microalgae.

The two to five-day old larvae did not have microalgae, the size of the shell is small, thin, smooth, the shell is formed through a transient amorphous calcium carbonate that acts as a precursor in the aragonite crystallization sequence [37, 38]. In addition, the time for colonization is still short.

In the case of larvae from 26 to 44 days, the highest species richness was reported, the shell is larger about 1 200 μm with greater ornamentation and with greater adhesion surface. Especially the 44-day-old larvae present a periostracum, outermost layer of the shell composed of an organic matrix [38, 39], which offers a better substrate, rich in proteins, which permit the growth of epizoic microalgae. In 28-day-old larvae, the number of spirals and shell ornamentation may be the factors that support the presence of *Haloleptolyngbya* sp., in addition to the mucus secreted by the microalgae themselves to adhere to the substrate. The 30 to 36-day old larvae showed a lower species richness and a dominance of *Cyclophora* sp. This diatom forms colonies in a zigzag shape, occupying the entire larva shell and preventing other microalgae from adhering. In several studies they have agreed to point out that gastropod shells are good microenvironments, due to their different structures and sizes [6]. Size does influence the colonization of epizoic microalgae, observing that the largest shells (*Bittium reticulatum*, *Gibbula adansoni*, *Columbella rustica* and *Clanculus cruciatus*) presented greater abundance and the small shells (*Alvania lineata*, *Nassarius incrassatus* and *Jujubinus striatus*) higher species richness, unlike what was found in this work.

Some microalgae produce allelopathic substances that inhibit the growth of others [40]. This could explain why some larvae had fewer epizoic microalgae than others, or for the fact that some had successfully colonized earlier and no longer left space for the colonization of more species. In addition to the changes in the abundances of the epizoic microalgae community, it is important to study the physical and chemical factors that influence their succession and to analyze whether *A. gigas* larvae fed on the epizoic microalgae reported. Epizoic microalgae associated with the velum of the larvae analyzed in this study were reported, *Cylindrotheca closterium*, *Hippodonta pseudacceptata* and *Cyclophora* sp.

The ecology studies of epizoic microalgae on the larvae of *A. gigas*, allows to know which species of phytoplankton or phytobenthos are present in the system where these larvae inhabit of the Mexican Caribbean. There are few studies focused on the study of diatoms and even less if they are found as epizoic microalgae. As the knowledge of the factors that regulate the competitive ability of the different epizoic microalgae species increases, the degree of interaction between them and their basibiont will also be understood.

Acknowledgements

We thank the National Council for Science and Technology (Conacyt) for grant N° 269167 and the project "The pink snail as an indicator of climate change in the

Caribbean: Ocean Warming and Acidification" (CB-2011-2101/181329). To the electron microscopy laboratory of the Mérida Unit of Cinvestav and to Ana Ruth Cristóbal Ramos and Dora Huerta for their work on electron microscopy.

Author details

Rocío Elizabeth Muciño-Márquez[1,3*], Dalila Aldana-Aranda[2], María Guadalupe Figueroa-Torres[3] and Oscar Hernández-Almeida[4]

1 Biological and Health Sciences, Autonomous Metropolitan University, Xochimilco Unit, Mexico

2 Cinvestav IPN, Mérida Unit Marine Ressources Departement, Mexico

3 Phycology Laboratory, Autonomous Metropolitan University, Xochimilco Unit, Mexico

4 Autonomous University of Nayarit, Mexico

*Address all correspondence to: mucinoelizabeth@gmail.com

References

[1] Wahl M, Marine epibiosis. I. Fouling and antifouling: some basic aspects. Marine Ecology Prog. Serie. 1989;58:175-189. DOI:10.3354/MEPS058175

[2] Lincoln RJ, Boxball GA, Clark PF, Diccionario de ecología, evolución y taxonomía. ed. México, D.F.. México: Fondo de Cultura Económica; 1995. 676 p. ISBN: 9786071600417

[3] Green J, Parasites and epibionts of Cladocera. Journal of Zoology. 1974;32:417-515. DOI: org/10.1111/j.1096-3642.1974.tb00031.x

[4] Ho J, Perkins PS, Symbionts of marine copepoda: An overview. Bulletin of Marine Science. 1985;37:586-598. URL: https://www.ingentaconnect.com/content/umrsmas/bullmar/1985/00000037/00000002/art00020#

[5] Gillan D, Cadée GC, Iron-encrusted diatom and bacteria epibiotic on *Hydrobia ulvae*. Journal of Sea Research. 2000;43:83-91. DOI: 10.1016/S1385-1101(99)00041-6

[6] D'Alelio D, Cante MT, Russo GF, Totti C, De Stefano M, Epizoic diatoms on gastropod shells. When substrate complexity selects for microcommunity complexity. In: Seckbach J, Dubinsky Z, editors. All Flesh Is Grass, Cellular Origin, Life in Extreme Habitats and Astrobiology, 1st ed. Springer: 2011. p. 16:345-364. DOI: 10.1007/978-90-481-9316-5_16

[7] Siqueiros-Beltrones DA, Ibarra-Obando S, Loya-Salinas, D, Una aproximación a la estructura florística de las diatomeas epifitas de Zostera marina y sus variaciones temporales en Bahía Falsa, San Quintín, B.C. Ciencias Marinas. 1985;11:3:69-88. DOI: org/10.7773/cm.v11i3.479

[8] Siqueiros-Beltrones DA, Serviere-Zaragoza E, Argumedo-Hernández U, Epiphytic diatoms of *Macrocystis pyrifera* (L.) C. Agardh from the Baja California Peninsula, Mexico. Oceánides, 2002;17:1: 31-39. URL: http://www.repositoriodigital.ipn.mx/handle/123456789/15141

[9] Siqueiros-Beltrones DA, Argumedo-Hernández U, Florística de diatomeas epifitas en laminales apicales de *Macrocystis pyrifera* (L.) C. Agardh. Oceánides, 2005;20:1,2:37-63. URL: https://oceanides.ipn.mx/index.php/cicimaroceanides/article/view/21

[10] Gárate-Lizárraga I, Muñetón-Gómez MS, Primer registro de la diatomea epibionte *Pseudohimantidium pacificum* y de otras asociaciones simbióticas en el Golfo de California. Acta Botánica Mexicana. 2009;88:33-47. URL: http://www.scielo.org.mx/pdf/abm/n88/n88a3.pdf

[11] Cupp EE, Allen WA, Plankton diatoms of the Gulf of California. Allan Hancock Pacific Expedition. 1938;35:61-99. URL: http://digitallibrary.usc.edu/cdm/ref/collection/p15799coll82/id/14856

[12] Moreno JL, Licea S, Santoyo H. Diatomeas del Golfo de California. 1 ed. Universidad Autónoma de Baja California Sur: SEP FOMES PROMARCO; 1996. 273 p.

[13] Cox E, Variation in patterns of valve morphogenesis between representatives of six biraphid diatom genera. Journal of Phycology. 1999;35:1297-1312. DOI: org/10.1046/j.1529-8817.1999.3561297.x

[14] Siqueiros-Beltrones DA, Diatomeas bentónicas de la Península de Baja California; diversidad y potencial ecológico. Oceánides. 2002;15:1:35-46. DOI: 10.7550/rmb.43748

[15] Komárek J, Anagnostidis K, Cyanoprokaryota 2. Teil/ 2nd Part: Oscillatoriales. In: Büdel B, Krienitz L, Gärtner G, Schagerl M, editors. Süsswasserflora von Mitteleuropa 19/2, Elsevier GmbH, München: 2005. p. 81-90. ISBN: 978-3-8274-1914-9

[16] Ashworth MP, Ruck EC, Lobban CS, Romanovicz DK, Theriot EC, A revision of the genus *Cyclophora* and description of *Astrosyne* gen. nov. (Bacillariophyta), two genera with the pyrenoids contained within pseudosepta. Phycologia. 2012;51:6:684-699. DOI: 10.2216/12-004.1

[17] Guiry MD, Guiry GM, AlgaeBase. World-wide electronic publication, National University of Ireland, Galway. 2020. http://www.algaebase.org; searched on 30 October 2020.

[18] Sullivan MJ, Similarity of an epiphytic and edaphic diatom community associated with *Spartina alterniflora*. Transactions of the American Microscopical Society. 1982;101:84-90. DOI: org/10.2307/3225573

[19] Clarke KR, Gorley RN. PRIMER v7: User Manual/Tutorial (Plymouth Routines in Multivariate Ecological Research). 2nd ed. PRIMER-E Ltd, Plymouth; 2006. 300 p. URL: http://updates.primer-e.com/primer7/manuals/User_manual_v7a.pdf

[20] Magurran AE. Diversidad ecológica y su medición. ed. Védra. Barcelona, España; 1989. 200 p. ISN: 9788487456008

[21] Sokal R, Rohlf FJ. Biometry. The principles and practice of statistics in biological research, 2nd ed. W.H. Freeman and Co., San Francisco; 1981. 859 p.

[22] Fee EJ, Drum RW, Diatoms epizoic on copepods parasitizing fishes in the Des Moines River, Iowa. The American Midland Naturalist. 1965;74:318-324.

[23] Hiromi J, Kadota S, Takano H, Diatom infestation of marine copepods (Review). Bulletin Tokai Regional Fisheries Research Laboratory. 1985;117:37-45.

[24] Winemiller KO, Winsborough BM, Occurrence of epizoic communities on the parasitic copepod *Lernaea carassii* (Lernaeidae). The Southwestern Naturalist. 1990;35:206-210. DOI: org/10.2307/3671543

[25] Gaiser EE, Bachmann RW, The ecology and taxonomy of epizoic diatoms on Cladocera. Limnology and Oceanography. 1993;38:628-637. DOI: org/10.4319/lo.1993.38.3.0628

[26] Round FE, Crawford RM, Mann DG. The diatoms. Biology, Morphology of the Genera. ed. Cambridge University Press, Cambridge. 1990. 747 p.

[27] Romagnoli T, Bavestrello G, Cucchiari EM, De Stefano M, Di Camillo CG, Pennesi C, Puce S, Totti C, Microalgal communities epibiontic on the marine hydroid *Eudendrium racemosum* in the Ligurian Sea during an annual cycle. Marine Biology. 2007;151:537-552. DOI: 10.1007/s00227-006-0487-x

[28] McClatchie S, Kawachi R, Dalley DE, Epizoic diatoms on the euphausid *Nyctiphanes australis*: consequences for gut-pigment analyses of whale krill. Marine Biology. 1990;104:227-232. DOI: 10.1007/BF01313262

[29] Hart TJ, On the diatoms of the skin film of whales and their possible bearing on the problems of whale movements. Discovery Reports, 1935;10:247-282. URL: https://www.icrwhale.org/pdf/SC01199-132.pdf

[30] Holmes RW, The morphology of diatoms epizoic on cetaceans and their transfer from *Cocconeis* to two new genera, *Bennettella* and *Epipellis*. British Phycological Journal. 1985;20:43-57. DOI: 10.1080/00071618500650061

[31] Croll DA, Holmes RW, A note on the occurrence of diatoms on the feathers of diving seabirds. Auk, 1982;99:765-766. URL: https://sora.unm.edu/sites/default/files/journals/auk/v099n04/p0765-p0766.pdf

[32] Holmes RW, Croll DA. Initial observations on the composition of dense diatom growths on the body feathers of three species of diving seabirds. In: D.G. Mann. ed. Proceedings of the 7th International Diatom Symposium. Koenigstein. O. Koeltz; 1984. p. 265-277.

[33] Cupul-Magaña FG, Cortés-Lara MC, Primer registro de epibiontes en ejemplares juveniles de *Crocodylus acutus* en el medio silvestre. Caldasia. 2005;27:1:147-149. URL: https://revistas.unal.edu.co/index.php/cal/article/view/39319

[34] Wahl M, Goecke F, Labes A, Dobretsov S, Weinberger F, The second skin: ecological role of epibiotic biofilms on marine organisms. Frontiers in Microbiology. 2012;292:1-21. DOI: 10.3389/fmicb.2012.00292

[35] Margalef R. Ecología. ed. Enciclopedia temática Planeta. Planeta. Barcelona. 1981. 252 p.

[36] Majewska R, Santoro M, Bolaños F, Chaves G, De Stefano M, Diatoms and Other Epibionts Associated with Olive Ridley (*Lepidochelys olivacea*) Sea Turtles from the Pacific Coast of Costa Rica. PLoS ONE. 2015;10:6:1-15. DOI: org/10.1371/journal.pone.0130351

[37] Brito-Manzano N, Aldana-Aranda D, Cruz-Lázaro EDela, Estrada-Botello MA, Organogenesis larvaria de *Strombus gigas* (Mesogastropoda: Strombidae) en el arrecife alacranes durante el periodo máximo de su época reproductiva. Universidad y Ciencia Trópico Húmedo. 2006;22:1:75-82. URL: https://www.redalyc.org/pdf/154/15402206.pdf

[38] Chávez-VillegasJF, Enríquez-DíazM, Aldana-Aranda D, Efecto de la temperatura y la acidificación en larvas de *Strombus gigas* (Mesogastropoda: Strombidae). Revista de Biología Tropical. 2017;65:2:505-515. DOI: 10.15517/RBT.V65I2.25504

[39] Enríquez-Diaz MR, Volland JM, Chávez-Villegas JF, Aldana-Aranda D, Gros O, Development of the planktotrophic veligers and plantigrades of *Strombus pugilis* (Gastropoda). Journal of Molluscan Studies. 2015;81:335-344. DOI: org/10.1093/mollus/eyv011

[40] Tanaka N, Asakawa A, Allelopathic effect of mucilage released from a brown alga *Sargassum horneri* on marine diatoms. Nippon Suisan Gakkaishi. 1988;54:1711-1714. URL: https://www.jstage.jst.go.jp/article/suisan1932/54/10/54_10_1711/_pdf

Chapter 5

The Use of Allelochemicals of Aquatic Macrophytes to Suppress the Development of Cyanobacterial "Blooms"

Evgeny Kurashov, Julia Krylova and Elena Protopopova

Abstract

Harmful algal "blooms", or HABs, is a hazardous natural phenomenon that often occurs under the influence of anthropogenic factors, for example, during the anthropogenic eutrophication of water bodies. An increase in the frequency and duration of cyanobacterial "blooms" carries a number of serious threats, including local and global degradation of water resources and the impact of cyanotoxins. There are various methods of fighting cyanobacterial "blooms" - physical, chemical, the use of bacterial preparations, etc. However, these methods are not effective enough and, most importantly, do not allow effectively solving the problem of suppressing HABs in water bodies without damage to other components of the aquatic ecosystem. Allelopathy is a natural phenomenon for both stimulatory and inhibitory effects of one plant upon another including microorganisms that resolves this problem. Allelochemicals of macrophytes can be considered as natural algaecides and become the basis of a nature-like convergent technology to suppress the development of plankton cyanobacteria and prevent HABs in water bodies. In our work, we used some allelochemicals of aquatic macrophytes to create a combined algicide of the new generation for suppressing the development of cyanobacteria. The effectiveness of suppressing cyanobacterial "blooms" is demonstrated by the example of field experiments with mesocosms and natural phytoplankton.

Keywords: harmful algal "blooms", phytoplankton, cyanobacteria, allelopathy, allelochemicals, field experiments, mesocosms

1. Introduction

Harmful algal "blooms", or HABs, is a hazardous natural phenomenon that often occurs under the influence of anthropogenic factors, for example, during the anthropogenic eutrophication of water bodies. An increase in the frequency and duration of cyanobacterial "blooms" carries many serious threats, including local and global degradation of water resources and the impact of cyanotoxins [1–3]. This problem is especially relevant and acute for millions of small reservoirs widely used for various types of water consumption: fisheries and aquaculture, water supply for various industries, including agricultural, drinking, and domestic water supply, recreational purposes, including sporting events. HABs occur when algae or cyanobacteria (most often they are) develop beyond measure and produce harmful effects on other hydrobionts, fish, aquatic and terrestrial animals, and birds as well as people [4, 5]. HABs disrupt the esthetics of water bodies and render the water unsuitable for various kinds of water uses. Economic damage due to HABs can be millions of dollars [6, 7].

Widespread HABs is a phenomenon to which special attention should be drawn since such "blooms" pose a number of serious threats, including local and global degradation of water resources and exposure to cyanotoxins [8–14].

Cyanobacterial "blooms" of water bodies are officially recognized as a global problem of modern ecology. Seasonal intense cyanobacterial "blooms" of reservoirs bring additional undesirable properties to natural and drinking water, such as a specific smell, taste, and the presence of toxins (microcystins). In some regions, the importance of this problem has been increasing recently [15]. The Working Group on the Evaluation of Carcinogenic Risks to Humans listed cyanotoxins as a carcinogenic substance harmful to humans [16].

The introduction of biotechnological methods into the practice of water body management that have maximum efficiency is one of the tasks of modern science. These include, first of all, the so-called convergent nature-like technologies, i.e. technologies that are based on any natural mechanisms causing this or that effect. These are precisely technologies that may be intended to ensure the sustainable development of modern countries [17–19].

Such technologies, aimed at managing the development of plankton communities in general and phytoplankton communities, in particular, may be based on such a phenomenon as allelopathy. This natural phenomenon can be very useful for effectively preventing and stopping the development of cyanobacterial "blooms" in water bodies [20–22]. Many existing methods of combating cyanobacteria [23] do not effectively solve the problem of "blooms" of water bodies without damage to other components of the ecosystem [3]. Usually, they are associated with serious adventitious effects on aquatic organisms and ecological systems [24].

At the same time, the application of the method of metabolic allelopathic control of HABs in water bodies during eutrophication is an effective and innovative solution to this problem. This approach preserves and restores water quality in water bodies, makes them suitable for multifunctional use, and natural allelochemicals (metabolites of macrophytes and their synthetic analogs) can be an effective alternative to existing algicides [20, 22, 25].

In reservoirs where macrophytes are developed (as a rule, at least 30% of the projective cover of the water area), water "bloom" is almost never observed. These circumstances are the causal basis for the development of nature-like technologies for the prevention and suppression of HABs with the help of new generation algicides based on allelochemical substances characteristic of aquatic macrophytes.

It has become apparent that metabolites-allelochemicals may be functioning in the processes of chemical suppressing of planktonic cyanobacteria in the aquatic ecosystems. However, data from field experiments are few concerning the effect of aquatic macrophyte allelochemicals on cyanobacteria, which is necessary for the development of nature-like technologies for preventing and suppressing cyanobacterial "blooms", and therefore they are the objects of "hottest" areas of research. Utilization of allelochemicals from aquatic macrophytes or using their synthetic analogs to inhibit cyanobacterial overgrowth is an environment-friendly technology for suppressing HABs.

Some reviews are focusing on the practice of the application of allelochemicals in agriculture [26, 27], but the field of using nature-like allelopathic technology to manage aquatic ecosystems is still poorly developed.

In the present study, we aimed to provide the information on the suppressing of cyanobacteria by macrophytes allelochemicals and the possibility to develop an algaecide of the new generation as a convergent nature-like technology for preventing and stopping the development of HABs in water bodies based on such a phenomenon as allelopathy.

2. Suppression of the development of cyanobacteria by aquatic macrophytes

Allelopathy as a natural phenomenon had been repeatedly recorded for a very long time in the 3rd century BC in ancient Chinese literature [28]. The term "allelopathy" was coined comparatively recently, in 1937 by Austrian plant physiologist Hans Molisch [29], who can be named as the father of allelopathy [30]. In general, we can consider allelopathy as an area of science, which investigates inhibitory or stimulatory biochemical interactions between the two plant/plant or plant/microorganism species.

The recent history of the study of low molecular weight organic compounds, which are small molecules (less than 900 amu) and constitute the low molecular weight metabolic profiles of organisms, should apparently begin with the discovery of the inhibitory effect of volatile plant excreta on microorganisms by Tokin Boris Petrovitch during the experimental work of 1928–1930 [31]. The research resulted in a number of publications, in one of which ("Bactericides of plant origin (phytoncides)") [32], the term "phytoncides" appeared. In the future, the doctrine of phytoncides was developed, which was reflected in the publication of several monographs. The history of research on phytoncides of aquatic and coastal plants began in the 40s of the XX century with the works of Gurevich Faiva Abramovich (1918–1992) [33], a student of B.P. Tokin. These studies ended in 1973 with the defense of a doctoral dissertation "Phytoncides of aquatic and coastal plants, their role in biocenoses" [34]. In particular, it was F.A. Gurevich who showed that the phytoncidal activity of aquatic plants is closely related to the macrophyte species and peculiarities of its development. He also showed that phytoncides are a very significant factor in the distribution of hydrobionts in a water body, including invertebrates.

At present, we can say that the macrophyte and algal allelopathy is paid much less attention than allelopathy in terrestrial ecosystems. Macrophytes and cyanobacteria are known to have an antagonistic relationship in different natural and experimental aquatic ecosystems [25, 35, 36].

It is a recognized fact that phytoplankton is poorly developed in macrophytic lakes. Even if we take into account the opinion that this is due to such factors as winning competition for nutrients and shading, then in the overwhelming number of cases, the main factor providing suppression of phytoplankton development is undoubtedly allelopathic suppression [37]. Apparently, the competition for nutrients cannot be recognized as a decisive factor in the outcome of the struggle between macrophytes and cyanobacteria, including considering that most aquatic macrophytes are rooted, and they usually obtain the main part of the necessary nutrients from the bottom sediments, which is characterized by high nutrient concentrations [38].

It is well known the phenomenon when shallow-water lakes can change their trophic status and the type of lake ecosystem, being either a pure water body with well-developed aquatic vegetation or a water body with low transparency, high turbidity, and intensive phytoplankton (mainly cyanobacteria) development. In other words, they can shift from one state to another [36, 39–43]. As this takes place, the mutual inhibitory allelopathic activities of macrophytes and phytoplankton may lead to the dominance of either macrophytes or phytoplankton [44].

We observed a similar effect in a floodplain lake with a changing trophic state in the Volga-Akhtuba interfluve, when cyanobacteria and macrophytes dominated in the same water body in different years [36]. Some evidence exists [45–48] that allelopathy is a factor affecting the development of phytoplankton (including cyanobacteria) in shallow lakes at the projective cover of macrophytes from 20 to 100%.

The importance of allelopathy as a powerful regulatory mechanism initiates a lot of studies devoted to the study of the inhibitory (sometimes stimulating) allelopathic effect of macrophytes on cyanobacteria and algae in aquatic ecosystems [49–58]. More than 60 species (67) of macrophytes are known to exhibit allelopathic activity against cyanobacteria. They are presented in **Table 1**.

According to the principle of allelopathic action, it is possible to prevent or mitigate the massive development of Cyanobacteria (blue-green algae), which leads to the HABs in water bodies. The implementation of this research direction promises huge benefits since it will solve the problem of the "blooms" of water bodies without negative consequences for other components of the ecosystem [20, 22, 25].

As follows from **Table 1**, data from laboratory studies, in general, prevail in the observation and proof of the effect of macrophyte allelopathy on cyanobacteria. These studies are based on laboratory-scale experiments using the co-cultures systems, adding plant extracts, or leachate collection. This state of affairs is associated with a more complex organization and interpretation of field studies. In this regard, data from field experiments and observations, for example with mesocosms, are of particular value. Numerous studies (including those included in **Table 1**) strongly suggest that allelopathy might thus be relevant in natural waters and suppress cyanobacteria and algae.

There are observations on the differentiation of the inhibitory effect of macrophytes on various species of cyanobacteria and algae. For example, it was concluded that the extracts, exudates, and live material of macroalgae *Chara australis* (Charophyta) exhibited strong inhibitory effects on the cyanobacterium *Trichormus variabilis (*formerly *Anabaena variabilis)*, but no effect was observed on the growth of the green alga *Scenedesmus quadricauda* [82].

The available data allow us to speak about the selective inhibition of various species of cyanobacteria by allelochemicals of various species of macrophytes. As a result, the allelopathic effect of macrophyte association on cyanobacteria (and all phytoplankton) seems to be stronger than the effect of one macrophyte species. This is evidenced by the fact that, as has been shown, the allelopathic effect of excretions of the association of macroalgae (*Chara hispida, C. baltica, C. vulgaris, Nitella hyaline)* and *Myriophyllum spicatum* is characterized by a significantly stronger effect than the effect of monoculture of macrophytes [83]. Such a combination of selective inhibition of macrophyte allelochemicals and a more strong impact of macrophyte assemblages toward the undesired cyanobacteria may be useful for biocontrol of HABs in water bodies as well as in aquaculture to remove harmful cyanobacteria and leave other algae to be used as food for hydrobionts and fish. The author [83] suggested that different allelochemicals produced by different macrophytes may exhibit a synergistic effect concerning cyanobacteria. It was also noted in [128] that different plants produce different types of allelochemicals and in different quantities. These summarized findings are therefore provided with more probability the basis for an effective strategy for reducing cyanobacterial biomass by introducing into water bodies with mixtures of submerged or floating native macrophytes for both restorations of aquatic ecosystems and mitigation of the HABs.

Species of macrophytes	Ecological form	Study scale	Cyanobacteria inhibited Study Scale	Source
Acorus tatarinowii, Acorus calamus, Acorus gramineus	EM	L	Cyanobacteriaas a whole	[59, 60]
Arundo donax	EM	L	*Microcystis aeruginosa*	[51, 57, 61–63]
Brasenia schreberi	FM	L	*Anabaena flos-aquae*	[64]
Cabomba caroliniana	SM	L	*Microcystis aeruginosa, Dolichospermum flosaquae* (formerly *Anabaena flos-aquae*), *Leptolyngbya tenuis* (formerly *Phormidium tenue*), Cyanobacteria as a whole	[65, 66]
Canna generalis	EM	L	*Microcystis aeruginosa*	[67]
Ceratophyllum demersum	SM	L, F	*Microcystis aeruginosa, Pseudanabaena limnetica* (formerly *Oscillatoria limnetica*), *Oscillatoriales. Anabaena sp., Trichormus variabilis* (formerly *Anabaena variabilis*), *Aphanizomenon flos-aquae, Synechococcus elongatus,* Cyanobacteria as a whole	[58, 68–78]; Our data
Chara aspera	SM	L	*Anabaena cylindrica, Anabaena torulosa, Anabaenopsis elenkinii, Microcystis aeruginosa, Microcystis flos-aqua, Synechococcus sp.,* Cyanobacteria as a whole	[37, 79–81]
Chara australis	SM	L	*Trichormus variabilis* (formerly *Anabaena variabilis*)	[82]
Chara baltica, C. canescens	SM	L	*Synechococcus sp.*	[81, 83]
Chara contraria	SM	L	*Anabaena cylindrica, Microcystis aeruginosa, Cylindrospermum sp.,* Cyanobacteria as a whole	[79]
Chara fragilis	SM	L*	*Oscillatoria limnetica,* Cyanobacteria as a whole	[71]
Chara globularis	SM	L	*Anabaena cylindrica, Anabaena torulosa, Anabaenopsis elenkinii, Planktothrix rubescens, Microcystis aeruginosa, Microcystis flos-aque, Cylindrospermum sp., Aphanizomenon flexuosum,* Cyanobacteria as a whole	[68, 72, 79, 84]
Chara hispida	SM	L, F	Cyanobacteria as a whole	[83, 85]
Chara rudis, Chara tomentosa, Chara delicatula	SM	L	*Anabaena cylindrica, Anabaena torulosa, Anabaenopsis elenkinii, Planktothrix agardhii, Planktothrix rubescens, Microcystis aeruginosa, Microcystis flos-aqua, Cylindrospermum sp., Aphanizomenon flexuosum,* Cyanobacteria as a whole	[79]
Chara vulgaris	SM	L, F	*Anabaena torulosa, Anabaenopsis elenkinii, Microcystis aeruginosa,* Cyanobacteria as a whole	[79, 83, 86, 87]
Cyperus alternifolius	EM	L	*Microcystis aeruginosa*	[67]

Species of macrophytes	Ecological form	Study scale	Cyanobacteria inhibited Study Scale	Source
Eichhornia crassipes	FM	L	*Microcystis aeruginosa, Microcystis sp., Raphidiopsis raciborskii* (formerly *Cylindrospermopsis raciborskii), Arthrospira platensis* (formerly *Spirulina platensis), Nostoc linckia* (formerly *Nostoc piscinale),* Cyanobacteria as a whole	[88–91]
Eleocharis acicularis	SM	L	Cyanobacteria as a whole	[66]
Eleocharis microcarpa	SM	L	*Anabaena flos-aquae, Oscillatoria tenuis*	[92, 93]
Elodea canadensis, Elodea nuttallii, Elodea sp.	SM	L, F	*Microcystis aeruginosa, Anabaena spp.,* Cyanobacteria as a whole	[35, 68, 78, 94, 95]
Hydrilla verticillata	SM	L	*Dactylococcopsis sp., Microcystis aeruginosa*	[56, 58, 96]
Egeria densa	SM	L	*Microcystis aeruginosa, Dolichospermum flosaquae* (formerly *Anabaena flos-aquae),*	[66]
Limnophila sessiliflora	SM		*Microcystis aeruginosa*	[66]
Myriophyllum aquaticum	SM	L	*Microcystis aeruginosa*	[97]
Myriophyllum brasiliense, Myriophyllum alterniflorum, Myriophyllum heterophyllum	SM	L	*Microcystis aeruginosa, Dolichospermum flosaquae* (formerly *Anabaena flos-aquae)*	[98]
Myriophyllum elatinoides	SM	L	*Microcystis aeruginosa*	[99]
Myriophyllum spicatum	SM	L, F	*Microcystis aeruginosa, Dolichospermum flosaquae* (formerly *Anabaena flos-aquae), Leptolyngbya tenuis* (formerly *Phormidium tenue);* Cyanobacteria as a whole	[54, 65, 71, 78, 83, 100–104]
Myriophyllum verticillatum	SM	L	Cyanobacteria as a whole	[105, 106]
Najas marina	SM	L	*Anabaena sp., Trichormus variabilis* (formerly *Anabaena variabilis), Synechococcus elongates,* Cyanobacteria as a whole	[74, 94]
Nasturtium officinale	EM	L	*Microcystis aeruginosa*	[107]
Nelumbo nucifera	FM	L, F	*Microcystis aeruginosa,* Cyanobacteria as a whole	[108, 109]
Nitella gracilis, Nitella opaca, Nitellopsis obtusa, Nitella hyaline, Nitella sp.,	SM	L, F	*Nitzschia palea, Anabaena cylindrica, Anabaena torulosa, Anabaenopsis elenkinii, Microcystis flos-aquae, Cylindrospermum sp., Aphanizomenon flexuosum,* Cyanobacteria as a whole	[68, 79, 83]
Nuphar lutea	FM	L, F	Cyanobacteria as a whole	[110]; Our data

Species of macrophytes	Ecological form	Study scale	Cyanobacteria inhibited Study Scale	Source
Nymphaea candida	FM	F	Cyanobacteria as a whole	Our data
Oryza sativa	EM		Cyanobacteria as a whole	[111]
Phragmites communis	EM	L	*Microcystis aeruginosa, Phormidium sp.*	[108, 112]
Pistia stratiotes	FM	L	*Synechococcus leopoliensis, Microcystis aeruginosa,*	[113–115]
Potamogeton crispus	SM	L, F	*Trichormus variabilis* (formerly *Anabaena variabilis*), Cyanobacteria as a whole	[82, 116, 117]
Potamogeton cristatus	SM	L	*Microcystis aeruginosa*	[58]
Potamogeton oxyphyllus	SM	L		[66]
Potamogeton lucens	SM	L, F	*Microcystis aeruginosa,* Cyanobacteria as a whole	[58, 71], Our data
Potamogeton maackianus	SM	L	*Microcystis aeruginosa*	[58, 118, 119]
Potamogeton malaianus	SM	L, F	*Microcystis aeruginosa, Oscillatoria sp.*	[118–120]
Potamogeton natans	SM	L, F	*Microcystis aeruginosa,* Cyanobacteria as a whole	[78], Our data
Potamogeton pectinatus	SM	L	*Microcystis aeruginos, Oscillatoria tenuis*	[76, 118, 121]
Ranunculus aquatilis	SM/FM	L	*Microcystis aeruginosa*	[107]
Ruppia maritima	SM	L	*Microcystis aeruginosa*	[122, 123]
Stratiotes aloides	FM	L, F	*Synechococcus elongatus, Microcystis aeruginosa,* Cyanobacteria as a whole	[49, 68, 71]
Typha latifolia, Typha minima, Typha angustata	EM	L	*Dolichospermum flosaquae* (formerly *Anabaena flos-aquae*), *Romeria leopoliensis* (formerly *Synechococcus leopoliensis*), *Microcystis aeruginosa*	[57, 124–126]
Vallisneria denseserrulata, Vallisneria spiralis, Vallisneria spinulosa	SM	L	*Microcystis aeruginosa*	[58, 66, 75, 127]

Table 1.
The number and relative content (% of total essential oil) of the fatty acids in some species of freshwater macrophytes and macroalgae from different water bodies.

Lombardo et al. [129] suggested that lake trophic state and extent of submerged vegetation coverage maybe the most important factors during formation in situ macrophyte–phytoplankton patterns at a large scale of natural water bodies. In this case, with a larger projective cover, a greater allelopathic effect will be achieved [45–48].

Not all macrophytes have the same allelopathic effect on cyanobacteria. Macrophytes that have the greatest suppressive effect on cyanobacteria (taking into account, among other things, information from **Table 1**) are such species and groups as *Cabomba caroliniana*, *Myriophyllum spicatum*, *Ceratophyllum demersum*, *Elodea canadensis*, *Nuphar lutea*, *Stratiotes aloides*, and family Characeae ([22, 36, 49, 65, 71, 103, 130], etc).

In the study [131], it was concluded that of all the 15 tested aquatic macrophytes, *Nymphaea odorata* and *Brasenia schreberi* have the highest allelopathic potential. However, this conclusion was obtained in experiments with lettuce sprouts, and not with cyanobacteria. These macrophytes inhibited 78% and 82% of lettuce seedling radicle growth and 98% and 68% of *L. minor* frond production respectively. Elakovich S. D. and Wooten J. W. [132] also reported that *Nuphar lutea* has high allelopathic activity.

Similar results were obtained with the macrophytes *Potamogeton maackianus*, *Potamogeton wrightii*, and *Potamogeton crispus*, which exhibited different inhibitory effects on the two species of algae [128]. There is a view that most allelochemicals are released during the early developmental stage of plants. It is assumed that during this period, plants are most dependent on stress conditions and competition with other surrounding plants for resources such as light, nutrients, and water [133]. However, in our studies, we found that the active synthesis of allelochemicals in aquatic macrophytes can continue even at later stages of plant development [22].

For the sake of completeness, it should be noted that some terrestrial plant materials (for example, barley straw) exhibit a strong allelopathic effect on cyanobacteria under certain conditions [134–136], which is no coincidence, since terrestrial plants also contain numerous allelochemicals [28]. It was shown in [137] that salcolin (two enantiomers that differ in their anti-cyanobacterial abilities) is the key allelochemical in barley straw's which exhibits an inhibitory effect on cyanobacteria and could be used as an agent in the control of cyanobacterial HABs. A review of typical terrestrial allelopathic plants with algistatic or algicidal effects is presented in [24].

3. Anti-cyanobacterial allelochemicals produced by aquatic macrophytes

Low-molecular-weight anti-cyanobacterial allelochemicals produced by aquatic macrophytes are very diverse. They belong to different classes of chemical compounds and are functionally diverse. Allelochemicals from the following groups of chemical compounds are the most important [22, 30, 55]: aldehydes, ketones, ethers, terpenes and terpenoids, phytoecdysteroids, fatty acids, sulfur-containing compounds, nitrogen-containing compounds, alcohols, lactones, polyacetylenes, quinines, phenolics, cinnamic acid and its derivatives, coumarins, flavonoids, tannins. These groups include hundreds of allelochemicals inhibiting cyanobacteria and algae [24], which should be discussed in detail in a special review.

These allelochemicals can be extracted from the plant biomass, but also their synthetic counterparts can be produced and used. This will reduce the consumption of natural plant resources. The effectiveness of synthetic allelochemicals can be similar to their natural counterparts. Thus, synthetic allelochemicals are a hopeful alternative to the use of natural metabolites-allelochemicals against HAB-forming cyanobacteria [20, 21].

Realizing that it is impossible to consider all groups of allelochemicals, here we will focus on considering only fatty acids and phenolic compounds as the most promising (in our opinion) for biotechnological use in the fight against HABs.

Studies of potential biological activities of major low molecular weight organic compounds of aquatic macrophytes using the QSAR method [138, 139] have shown that fatty acids and gallic acid are characterized by various types of bioactivity with the highest probability of manifestation (Pa > 0.9) that can induce cyanobacteria growth suppression. Further studies based on the results obtained suggest clarifying experimental studies of the reaction of various species of cyanobacteria to the effects of selected allelochemicals.

As it was received in laboratory experiments conducted with fatty acids for their effect on the cyanobacteria *Synechocystis aquatilis* and *Aphanizomenon flos-aquae*, and which are described in detail in [140], selected allelochemicals (linoleic, heptanoic, octanoic, tetradecanoic, hexadecanoic, and gallic acids) possess inhibitory allelopathic activity against cyanobacteria. However, their inhibitory effect was different. The highest values of the Suppression index (SI, defined as the cyanobacterial density in control divided by the cyanobacterial density in an experiment with allelochemicals) (SI > 10) were recorded (in ascending order) for hexadecanoic, linoleic, tetradecanoic, gallic acids, and a mixture of four allelochemicals (heptanoic, octanoic, tetradecanoic and gallic acids).

The highest SI values for Synechocystis aquatilis were obtained when the culture of cyanobacteria was exposed to gallic acid (SI = 30) and a mixture of heptanoic, octanoic, tetradecanoic, and gallic acids (SI = 35.3). *Aphanizomenon flos-aquae* was found to be more sensitive to the effect of the given mixture of allelochemicals. SI for it on the 23rd day of the experiment was 17495 [140].

In works [141, 142] problems have been raised concerning effective algal inhibitors and control HABs. To address these issues, the authors suggested using unsaturated fatty acid (linoleic acid) in conjunction with alginate – chitosan microcapsule technology. They demonstrated that the linoleic acid microsphere had good encapsulation efficiency and release property. Besides, linoleic acid sustained-released microspheres could inhibit *Microcystis aeruginosa* (Cyanobacteria) growth to the non-growth state, and thus linoleic acid microsphere may be used as a potential candidate for HABs control.

Studies on the use of microgranules saturated with an allelochemical or a combination of allelochemicals (for example, a combination of fatty acids and phenolic compounds) to suppress cyanobacteria look very promising. The inhibitory agent, gradually releasing from the microgranules, prolongs its allelopathic effect on cyanobacteria. A sustained-release time of allelochemicals can range from 40 to 120 days [142–144]. A review of the studies carried out in this direction is presented in [128]. Results obtained in different investigations open up new promising areas for scientific research and practical use of allelochemicals of aquatic macrophytes.

According to results received in [112], nonanoic acid can inhibit the growth of cyanobacteria *Leptolyngbya tenuis* (formerly *Phormidium tenue)* and *M. aeruginosa*, whereas, no inhibitory effects of stearic, and palmitic acids was found.

In earlier works [113, 125], it was also found, that three fatty acids (α – linolenic, linoleic, and an unidentified C8:2) inhibited cyanobacteria (particularly T 625 *Romeria leopoliensis* (formerly *Synechococcus leopoliensis)* and T 1444 *Dolichospermum flosaquae* (formerly *Anabaena flosaquae*)).

The essential oil of some allelopathic plants (*Potamogeton cristatus, Potamogeton maackianus, Potamogeton lucens, Vallisneria spinulosa, Ceratophyllum demersum, and Hydrilla verticillata*) was demonstrated to inhibited *Microcystis aeruginosa*, during which fatty acids constituted an important part of the essential oils isolated.

Recently, Wang et al. [95] reported the inhibitory effects of some fatty acids on *Microcystis aeruginosa*. The authors stated that pentadecanoic acid, linoleic acid, alpha-linolenic acid, and stearic acid were the most potent allelochemicals from *Elodea nuttallii* along with dihydroactinidiolide and beta-ionone.

We showed [140] that such plants as *Potamogeton natans*, *Nuphar lutea*, *Nymphaea alba*, *Myriophyllum spicatum*, *Persicaria amphibia* are the most active producers of allelochemical fatty acids, and therefore they can have a significant allelopathic effect on cyanobacteria and phytoplankton in total. In these plants, the proportion of fatty acids in the content of volatile organic compounds can exceed 60–70%.

Our studies of the metabolome of *Potamogeton perfoliatus* from different habitats in Lake Ladoga show that the abundance of cyanobacteria in the associations of this macrophyte depends on the content of carboxylic acids in a given plant (**Figure 1**).

The study by Gao et al. [145] demonstrates that nonanoic acid may be involved in synergistic interactions with other allelochemicals, demonstrating a stronger allelopathic effect against Microcystis aeruginosa.

Similar results were obtained for octadecanoic acid [146], which may participate in synergistic, antagonistic, and additive allelopathic interactions. These findings led to the conclusion that joint effects of different allelochemicals depend on various factors such as the chemicals used, their respective proportions, the total concentration of the mixture, and the receptor species [146].

In addition to fatty acids, among allelochemicals, special attention should be paid to phenolic compounds.

As early as in 1981 [100], the results were published, which demonstrated that phenolic compounds extracted from *Myriophyllum spicatum* exhibit algicidal activity against cultured algae and natural phytoplankton assemblages. Later, it was found that such aquatic macrophytes as representatives of the genus Myriophyllum are able to excrete polyphenol-like allelochemicals to inhibit the growth of green algae and cyanobacteria [98]. A number of identified polyphenols (ellagic, gallic, pyrogallic, and catechin) and fatty acids (hexadecanoic acid, stearic acid, α-linolenic acid) were shown to significantly suppress the development of HAB-forming cyanobacteria species [147, 148].

Additionally, a study [78] has revealed that the major allelochemicals identified in tested macrophyte ethyl acetate extract of *Nasturtium officinale* included quercetin, tannic acid, and gallic acid. Also, findings are the combinations of different types of polyphenols, such as pyrogallic acid, gallic acid, and ellagic acid may have

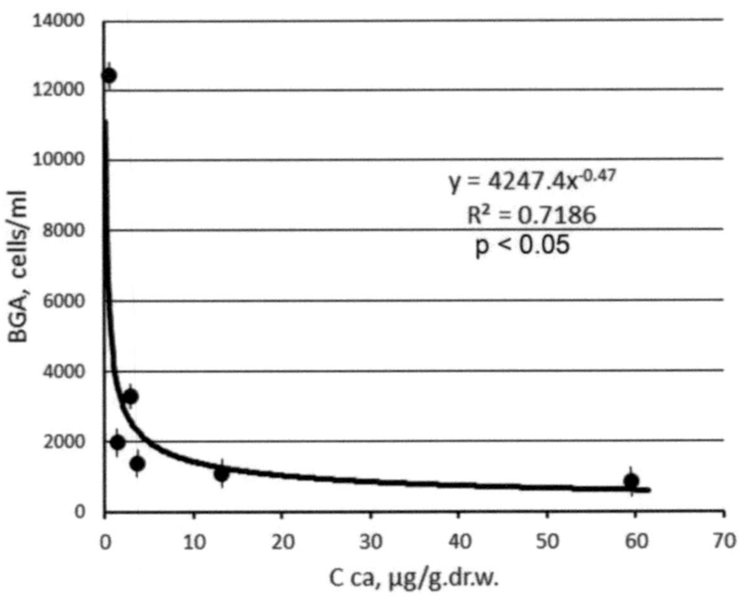

Figure 1.
Dependence of the concentration of cyanobacteria (BGA, cells/ml) on the concentration of fatty acids (Cca, µg/g.dr.w.) in Potamogeton perfoliatus *in Lake Ladoga.*

an additive or synergistic effect on cyanobacterium *Microcystis aeruginosa* and the joint action of phenolic allelochemicals may be an important allelopathic pattern of submerged macrophytes to inhibit the growth of HAB-forming cyanobacteria in natural aquatic ecosystems [53, 146, 148–150].

In a study [54] during the investigation of contributions of five allelochemicals, (+) catechin, eugeniin, and ellagic, gallic, and pyrogallic acid, in the allelopathic effects of *Myriophyllum spicatum* on the cyanobacterium *M. aeruginosa* it was observed that these compounds, on average, may provide up to 50% of the allelopathic effects of *M. spicatum*. According to results received in [112], four phenols (sinapic, syringic, caffeic, and gallic acids) inhibited the growth of cyanobacteria *Leptolyngbya tenuis* (formerly *Phormidium tenue*) and *M. aeruginosa*. The inhibitory effect of pyrogallic acid and gallic acid produced by *M. spicatum* in relation with cyanobacteria was also demonstrated in [53, 114].

It is beyond question that there is a huge amount of scientific material regarding the allelopathic properties of fatty acids and gallic acid ([52, 54, 56, 67, 88, 103, 112, 113, 118, 119, 124–126, 146, 148, 151–166], etc.). This circumstance gives every reason to use them to create a new generation of algicides based on allelochemical substances of aquatic macrophytes. The use of this information, as well as the results of our researches [36, 138, 140], formed a prerequisite for the development of a new generation algicide based on allelochemicals of aquatic macrophytes against cyanobacteria. It is precisely fatty acids (heptanoic, octanoic, tetradecanoic acids) and gallic acid that were included in its composition [167].

4. Mesocosm study of the effects of allelochemicals on cyanobacteria

Evidence of suppression of the development of phytoplankton, including planktonic cyanobacteria, in real natural conditions by traditional observations, even in the most obvious cases [36], is nevertheless indirect and often contradictory [48, 168]. Taking this into account, the way of assessing the effect of allelochemicals on cyanobacteria in experiments with mesocosms in natural conditions is more promising and makes it possible to obtain results corresponding to natural aquatic ecosystems.

A good example is a field study by Hilt et al. [169] in which the authors found an allelopathic effect of the macrophyte *Myriophyllum verticillatum* on natural phytoplankton (including cyanobacteria) in Lake Krumme Lake (Berlin, Germany). In a mesocosm study [170] in Laguna Blanca lake in Manantiales (Maldo-nado, Uruguay) it was observed that macrophytes species (Egeria densa and *Potamogeton illinoensis*) seem to exert strong biological effects on phytoplankton biomass, and they are able to keep phytoplankton biomass low through allelopathic influence, even in the absence of zooplankton grazing.

In another mesocosm study [171], similar results were obtained, demonstrating that another species of the genus Myriophyllum (*Myriophyllum spicatum*) under conditions of 85 l mesocosms during 13 days of exposure had an only short-term inhibitory effect on total phytoplankton and green algae, whereas consistent negative effects (allelopathic) were detected concerning *M. aeruginosa*.

After the development of an algicide containing fatty acids (heptanoic, octanoic, tetradecanoic acids) and gallic acid, the rationale for the use of which is presented in detail in [140], we conducted the first experiments with this algicide with natural phytoplankton communities under conditions mesocosms.

In the field experiments, mesocosms with a volume of 700 liters were used. The experiments were carried out on two ponds on the territory of St. Petersburg (Russia): at Pulkovo Pond (pond 1; coordinates 59.835899, 30.328642) and Aviator's

Pond (Pond 2; coordinates 59.868343, 30.300443). The depth of the ponds at the location of the experiments was about 3 m. The mesocosms were filled with water from the pond, then algicide was added to them in an amount so that its concentration in the water of the mesocosms was 1 mg/l.

In Pulkovo Pond, the experiment was carried out from June 25 to July 5, 2019. In the Aviatorov Pond, the experiment was carried out from July 2 to July 16, 2019. The temperature and light conditions in the mesocosms corresponded to those in the water of the pond outside the mesocosms. The change in water temperature in the surface layer of the studied ponds is shown in **Figure 2**.

The results of the algicide impact on the phytoplankton of pond 1 are shown in **Figures 3–6**.

As can be seen from **Figure 3**, in the water of pond 1, both the abundance and the biomass of all phytoplankton increased during the experiment. At the same time, this was not observed in the mesocosm. In the first three days, a decrease in phytoplankton biomass without a change in its abundance occurred. Subsequently, the abundance and biomass of phytoplankton in the mesocosm remained approximately at the same level as they grew in the pond. By the end of the experiment (on the 11th day), the phytoplankton biomass in the pond exceeded that in the mesocosm by about 5 times, and the abundance - by almost 12 times. The greatest differences were observed on the 8th day of the experiment; the difference in biomass and abundance was 7 and 20 times, respectively. Thus, the action of an algicide based on fatty acids and gallic acid inhibited the growth of phytoplankton.

The data of phytoplankton analysis are confirmed by the data on the measurement of optical density in the pond and the mesocosm (**Figure 4**). By the end of the experiment, an increase in optical density in the pond and a significant decrease in optical density in the mesocosm were observed (**Figure 4**). By the end of the experiment, the difference was about 2.3 times. This was also noticeable visually: the water in the mesocosm was more transparent than the water in the pond surrounding the mesocosm (**Figure 5**).

It is interesting to trace how the quantitative indicators of cyanobacteria in the pond and the mesocosm changed. *Dolichospermum solitarium* (formerly *Anabaena solitaria*) was the dominant cyanobacterial species in the pond (and at the beginning of the experiment in the mesocosm). This species belongs to cyanobacteria

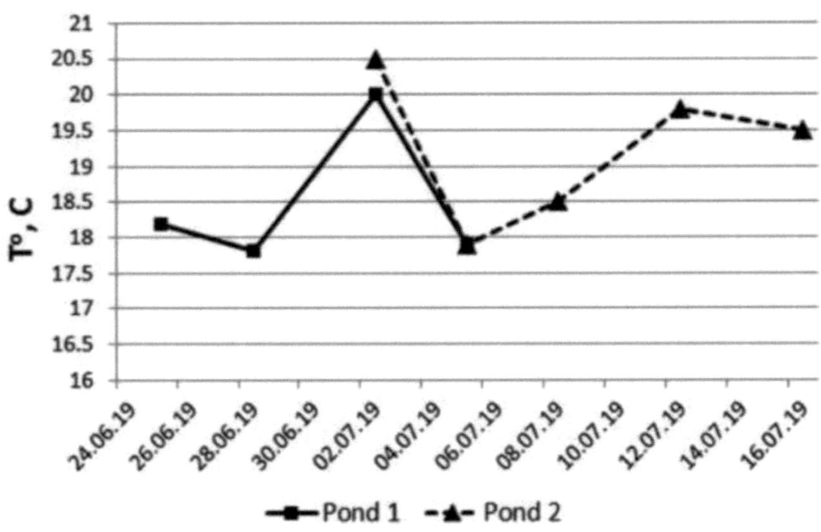

Figure 2.
Change in water temperature (° C) in the surface layer of the investigated ponds.

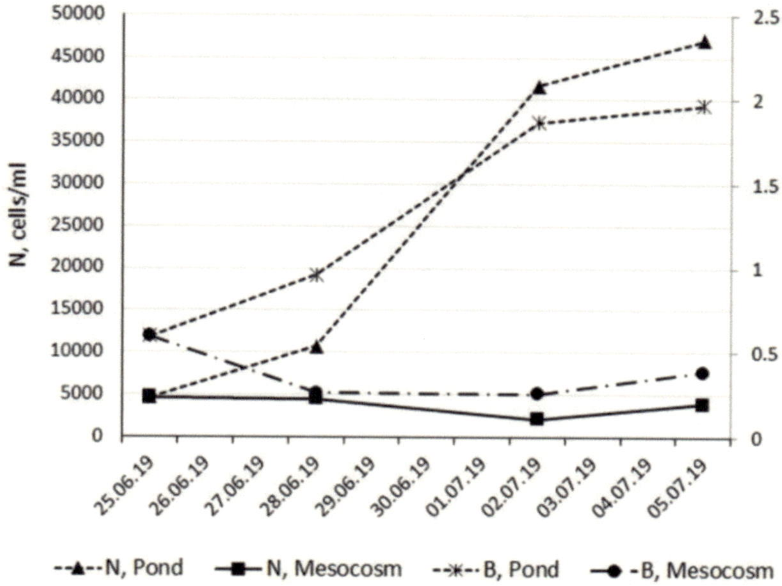

Figure 3.
Changes in the abundance and biomass of total phytoplankton in pond 1 and the mesocosm under the influence of algicide with a concentration of 1 mg/l.

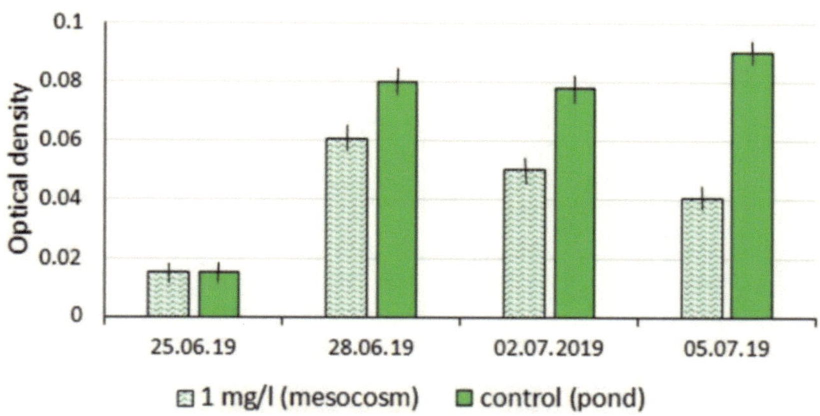

Figure 4.
Change in the optical density of the water mass in pond 1 and the mesocosm when exposed to algicide with a concentration of 1 mg/l.

capable of causing the phenomenon of HABs [172]. A decrease in both the number and biomass of cyanobacteria both in the pond and in the mesocosm was observed on the third day of the experiment. Moreover, in the mesocosm, this decrease was more pronounced. Subsequently, an increase in the number and biomass of cyanobacteria both in the pond and in the mesocosm was observed. However, it was more intense in the pond. By the end of the experiment (on the 11th day), the biomass of cyanobacteria in the pond exceeded that in the mesocosm by about 2.5 times, and the number - by 1.5 times. The greatest differences were observed on the 8th day of the experiment, the difference in biomass and abundance was 4.4 and 39 times, respectively. At the end of the experiment, the same species *Dolichospermum solitarium* remained the dominant species in the composition of cyanobacteria. At the same time, *Cuspidothrix ussaczevii (formerly Aphanizomenon elenkinii)* began to dominate in the mesocosm among cyanobacteria. This species is also included in

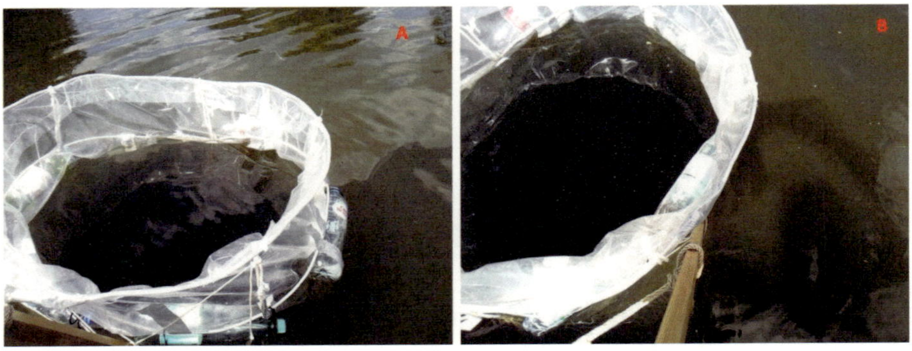

Figure 5.
The contrast in the state of water mass in pond 1 and mesocosm 4 (a) and 11 (B) days after exposure to algicide.

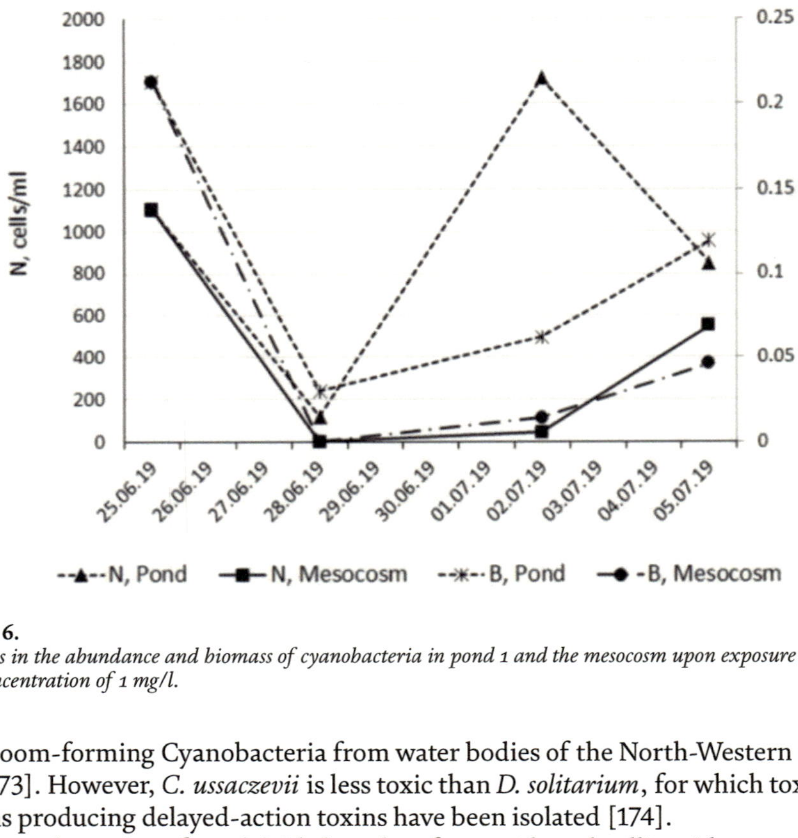

Figure 6.
Changes in the abundance and biomass of cyanobacteria in pond 1 and the mesocosm upon exposure to algicide at a concentration of 1 mg/l.

the bloom-forming Cyanobacteria from water bodies of the North-Western Russia list [173]. However, *C. ussaczevii* is less toxic than *D. solitarium*, for which toxigenic strains producing delayed-action toxins have been isolated [174].

Thus, the action of an algicide based on fatty acids and gallic acid prevented the growth of the number of cyanobacteria and changed their species structure.

In pond 2, the beginning of the experiment coincided with an intense cyanobacterial "bloom" (**Figure** 7), while their biomass was more than 55 mg/l. At the same time, in the surface layer of the pond, the maximum water temperature (20.5°C) for the entire duration of the experiment was noted (**Figure 2**). The cyanobacteria *Aphanizomenon flos-aquae*, *C. ussaczevii*, and *Dolichospermum affine* (formerly *Anabaena affinis)* dominated in phytoplankton. *Aphanizomenon flos-aquae* is one of the most widespread species that form HABs in ponds and lakes in Northwest Russia [173]. The species is capable of synthesizing dangerous (including for humans) toxins [173]. *Cuspidothrix ussaczevii* also often causes water

Figure 7.
Cyanobacterial HAB in pond 2 and water-filled mesocosm on July 2, 2019.

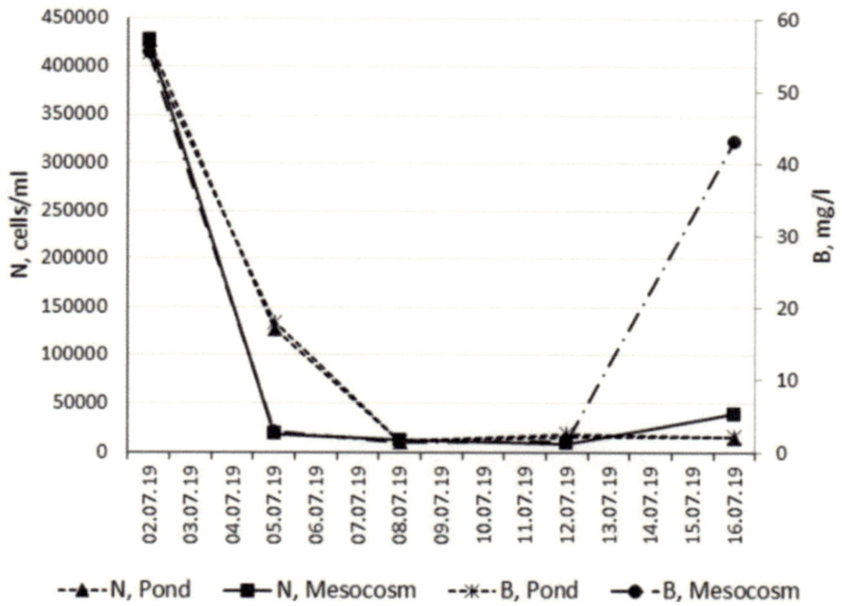

Figure 8.
Changes in the abundance and biomass of total phytoplankton in pond 2 and the mesocosm under the influence of algicide with a concentration of 1 mg/l.

"bloom" in water bodies of St. Petersburg and the Leningrad Region, being the dominant or subdominant in bloom-forming cyanobacteria [173].

By the fourth day of the experiment, the water temperature in the pond dropped to about 18°C. This led to a decrease in the number and biomass of cyanobacteria, apparently, mainly due to their sinking into the lower layers of the reservoir. However, an even greater decrease in the development of cyanobacteria was observed in the mesocosm, in which cyanobacteria could not sink so deeply (**Figure 8**). This is also confirmed by data on the optical density of water in the pond and in the mesocosm, where a more significant decrease was noted (**Figure 9**). Subsequently, the optical density slightly decreased to approximately the same level in the pond and mesocosm and almost did not change in the pond and mesocosm.

At the same time, the control of the development of cyanobacteria from pond 2 in the laboratory, where there was no decrease in temperature, showed their significant growth in the control. With that, under the influence of allelochemicals, significant suppression of plankton growth was observed, recorded by optical density (**Figure 10**).

By the 8th day of the experiment, a further decrease in the optical density of plankton under the influence of algicide was noted in the laboratory. At the same time, a decrease in optical density and the control was observed, obviously, due to the inability of natural plankton to laboratory conditions (the experiment was carried out in 0.5-liter jars).

By July 8, the species of cyanobacteria *Aphanizomenon flos-aquae* and *Cuspidothrix ussaczevii* in the mesocosm dropped out of the dominant composition, although they continued to dominate in the pond water. As our laboratory experiments with this algicide have shown [140], this species of cyanobacteria was especially sensitive to the used mixture of allelochemicals. So, a complete suppression of the development of the culture of *Aphanizomenon flos-aquae* was observed in the experiment with the combined effect of heptanoic, octanoic, tetradecanoic, and gallic acids at various concentrations (0.1, 1, and 10 mg/l).

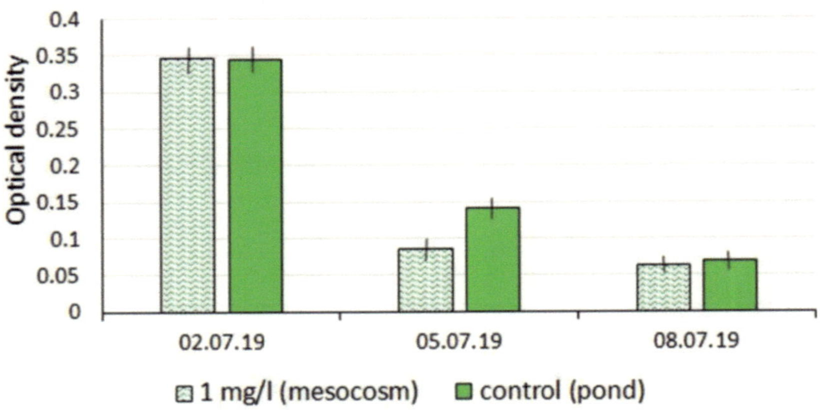

Figure 9.
Change in the optical density of the water mass in pond 2 and the mesocosm when exposed to algicide with a concentration of 1 mg/l.

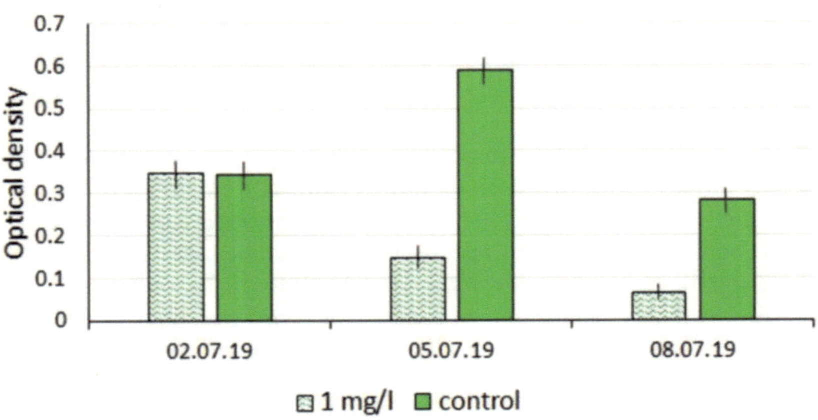

Figure 10.
Change in the optical density of the water mass in pond 2 and the mesocosm when exposed to algicide with a concentration of 1 mg/l during exposure in the laboratory.

In the last phase of the experiment (from July 12), representatives of Cryptophyta - *Cryptomonas sp.*, *Komma caudata* (formerly *Chroomonas acuta*) dominated the pond in the composition of phytoplankton (**Figure 11**). Among the cyanobacteria, *Aphanizomenon flos-aquae* and *Aphanocapsa conferta* dominated. In the mesocosm at this time (especially toward the end of the experiment) cryptophyte algae (98% of the total phytoplankton biomass) with the dominant *Cryptomonas sp*. reached a very high development (with biomass of more than 42 mg/l) (**Figure 11**). Cyanobacteria were represented by the species *Dolichospermum affine*, *Aphanocapsa conferta* with very little quantitative development.

It is noteworthy that by the end of the experiment in the mesocosm, the total phytoplankton biomass returned to almost the same high values as at the beginning of the

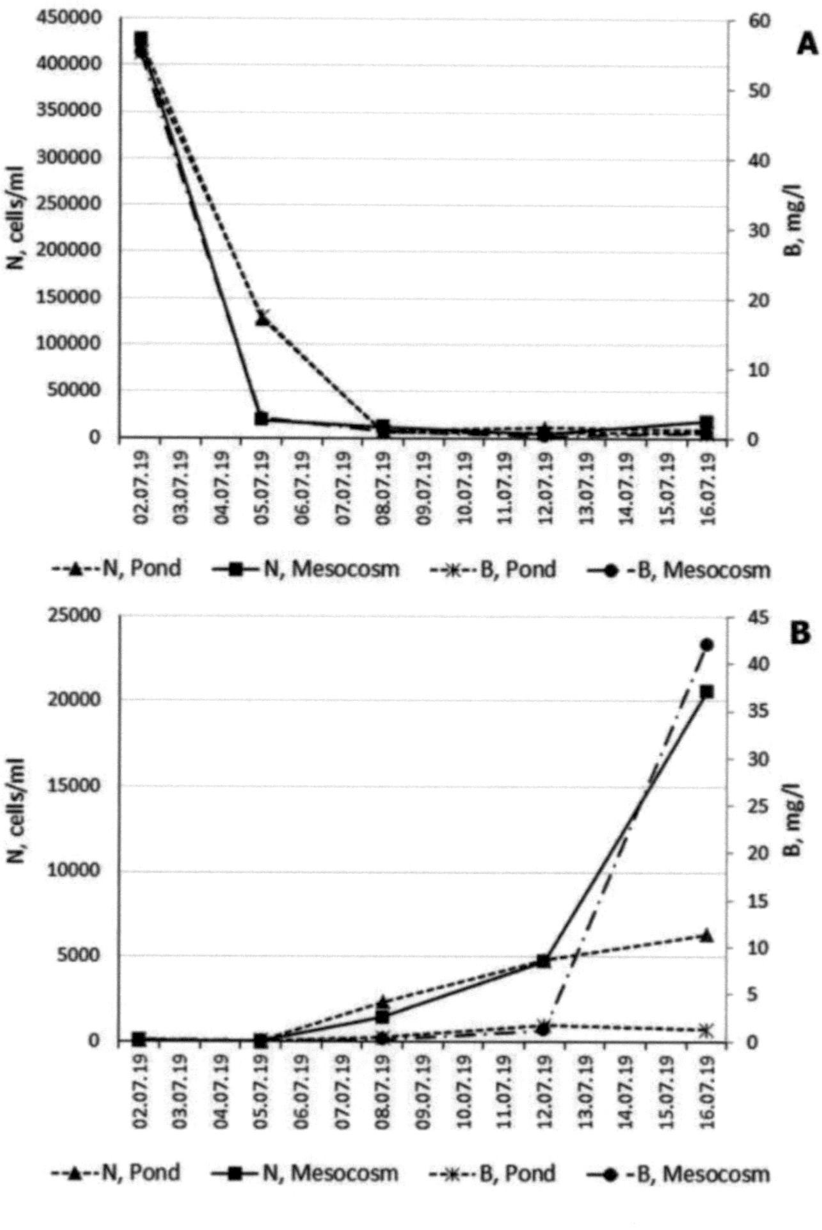

Figure 11.
Changes in the abundance and biomass of cyanobacteria (a) and Cryptophyta (B) in pond 2 and the mesocosm under the influence of algicide at a concentration of 1 mg/l.

experiment. However, if at the beginning of the experiment cyanobacteria prevailed (about 99% of the total biomass of phytoplankton), then by the end of the experiment cryptophyte algae accounted for more than 98% of the biomass of phytoplankton. *Cryptomonas sp.* dominated among cryptophyte algae. That is, the replacement of dangerous toxicogenic species of cyanobacteria with cryptophyte algae occurred, which can be consumed by aquatic organisms and which are safe for other organisms, including humans. If we project this result to entire aquatic ecosystems, then we can get a very beneficial ecosystem effect, expressed in the suppression of HABs and the development of algae, whose production can be consumed, for example by zooplankton and planktivorous fish.

Thus, the main results of the experiments carried out on the effect of an algicide of four allelochemical components (heptanoic, octanoic, tetradecanoic, and gallic acids) on the phytoplankton of natural water bodies can be considered the following results, indicating that allelochemical substances of aquatic macrophytes: 1) are able to effectively reduce phytoplankton development and suppress even intense HABs; 2) may lead to the replacement of dangerous cyanobacteria in phytoplankton with safe algae, whose production can be used in the food chains of aquatic organisms.

5. Conclusions and perspectives

In this way, available data show that the use of allelochemicals from aquatic macrophytes to inhibit cyanobacterial overgrowth is an environment-friendly and perspective technology for suppressing HABs. Allelochemicals can be considered as natural algaecides and become the basis of a nature-like convergent technology to mitigate the development of plankton cyanobacteria and prevent HABs in water bodies.

One can quite agree with the conclusion of work [24] that allelopathy is a promising strategy to control HABs as the effectiveness of allelochemicals on inhibiting microalgae cells has been discovered, investigated, and confirmed in many works and for many years [175]. However, there are several problems that must be investigated in order to understand what determines the strength of the manifestation of the allelopathic effect. One of these problems is undoubtedly the action of various environmental factors.

Another problem is the resistance of allelochemicals in the aquatic environment and their chemical or biochemical (under the influence of bacteria) changes [26, 74, 168, 176]. In this regard, very promising are works in which systems are being developed that allow dosing and prolonging the release of allelochemicals into the aquatic environment [141–143].

The development and research of allelopathy and its application for suppressing the HABs are striving toward a future for sustainable, rational, and effective using the water resources worldwide. The algicides of the new generation developed based on the phenomenon of allelopathy can definitely reduce the amount of synthetic algicides and herbicides used.

While allelochemicals have shown growth inhibition of planktonic cyanobacteria, there is still insufficient knowledge of the impact on various species of cyanobacteria (especially their action in real aquatic ecosystems), the influence of various factors on the action of allelochemicals, and the molecular mechanisms of their action. These gaps may limit their use as conventional biotechnology for the mitigation and prevention of HABs in aquatic ecosystems.

All the laboratory studies can propose only the potential for allelopathy of macrophytes metabolites toward cyanobacteria, its real use as biotechnology for the management of planktonic communities and HABs will be possible only after convincing field studies using mesocosms and entire ecosystems.

In addition, if we are to understand more about the mechanisms of allelochemicals actions that cyanobacterial cells respond to, more cognizance needs to be taken of the molecular peculiarities of interactions between allelochemicals and cyanobacterial cells.

Acknowledgements

The work was performed within the framework of the state task of the Russian Academy of Sciences on topic 0154-2019-0002. The authors thank Dr. Alexandr Rusanov and Mr. Denis Bardinskij as well as Ms. Elena Fisak for their kind help in the field experiments.

Conflict of interest

The authors declare that there is no conflict of interest.

Author details

Evgeny Kurashov[1,2]*, Julia Krylova[2] and Elena Protopopova[1]

1 Institute of Limnology, A Separate Subdivision of the St. Petersburg Federal Research Center of the Russian Academy of Sciences, Saint Petersburg, Russia

2 Saint-Petersburg Branch of the Federal State Budgetary Scientific Institution "All-Russian Research Institute of Fisheries and Oceanography" ("GosNiorch" by L.S. Berg), Saint Petersburg, Russia

*Address all correspondence to: evgeny_kurashov@mail.ru

References

[1] Anderson D. HABs in a changing world: a perspective on harmful algal blooms, their impacts, and research and management in a dynamic era of climactic and environmental change. Harmful Algae. 2012 (2012). 2014;2012:3-17

[2] Šulčius S, Montvydienė D, Mazur-Marzec H, Kasperovičienė J, Rulevičius R, Cibulskaitė Ž. The profound effect of harmful cyanobacterial blooms: From food-web and management perspectives. Science of The Total Environment. 2017;609:1443-1450. DOI:10.1016/j.scitotenv.2017.07.253

[3] Huisman J, Codd GA, Paerl HW, Ibelings BW, Verspagen JMH, Visser PM. Cyanobacterial blooms. Nat Rev Microbiol. 2018;16:471-483. DOI:10.1038/s41579-018-0040-1

[4] Hallegraeff GM. A review of harmful algal blooms and their apparent global increase. Phycologia. 1993;32(2):79-99. DOI:10.2216/i0031-8884-32-2-79.1

[5] Amorim CA, Moura AN. Ecological impacts of freshwater algal blooms on water quality, plankton biodiversity, structure, and ecosystem functioning. Science of The Total Environment, Available online 17 November 2020. DOI: 10.1016/j.scitotenv.2020.143605

[6] Hoagland P, Anderson D M, Kaoru Y, White AW. The Economic Effects of Harmful Algal Blooms in the United States: Estimates, Assessment Issues, and Information Needs. Estuaries. 2002;25(4b):819-837

[7] Wolf D, Georgic W, Klaiber HA. Reeling in the damages: Harmful algal blooms' impact on Lake Erie's recreational fishing industry. Journal of Environmental Management. 2017;199:148-157. DOI:10.1016/j.jenvman.2017.05.031

[8] Brock TD. A Eutrophic Lake: Lake Mendota, Wisconsin. Ecological Studies. Springer-Verlag New York Inc. 1985;55. 308 p. DOI: 10.1007/978-1-4419-8700-6

[9] Landner L, Wahlgren U. Eutrophication of lakes and reservoirs in warm climates. Copenhagen; World Health Organization. Regional Office for Europe; 1988. 121 p.

[10] Harper D. Eutrophication of freshwaters, principles, problems and restoration. Chapman and Hall, London; 1992. 327 p.

[11] Lam AKY, Prepas LEE, Spink D, Hrudey SE. Chemical control of hepatotoxic phytoplankton blooms: Implications for human health. Water Res. 1995;29(8):1845-1854

[12] Ansari AA, Gill SS, Lanza GR, Rast W, editors. Eutrophication: Causes, Consequences and Control. Springer, Dordrecht; 2011. 394 p. DOI: 10.1007/978-90-481-9625-8

[13] Chislock MF, Doster E, Zitomer RA & Wilson AE. Eutrophication: Causes, Consequences, and Controls in Aquatic Ecosystems. Nature Education Knowledge. 2013;4(4):10

[14] Yanagi T. Eutrophication and Oligotrophication in Japanese Estuaries: A Synthesis. In: Yanagi T, editor. Eutrophication and Oligotrophication in Japanese Estuaries. Estuaries of the World. Springer, Dordrecht. 2015. p. 1-4. DOI:10.1007/978-94-017-9915-7_1

[15] Rivasa EJG, Pérezb GR, Tundisic JG, Vammend K, Örmecie B, Fordef M. Eutrophication: A growing problem in the Americas and the Caribbean. Braz. J. Biol. 2020;80(3). DOI:10.1590/1519-6984.200001

[16] IARC Monographs on the Evaluation of Carcinogenic Risks to Humans.

Ingested Nitrate and Nitrite, and Cyanobacterial Peptide Toxins. Vol.94. Lyon, France, 450 p.

[17] Koval'chuk MV, Naraikin OS. Nature-Like Technologies—New Capacities and New Challenges. Ind. Bezop. 2017;22(3-4):118-119

[18] Zhironkin S, Demchenko S, Kayachev G, Taran E, Zhironkina O. Convergent and Nature-Like Technologies as the Basis for Sustainable Development in the 21st Century. IVth International Innovative Mining Symposium. E3S Web of Conferences 105, 03008. 2019. DOI:10.1051/e3sconf/201910503008

[19] Nature-like and Convergent Technologies Driving the Fourth Industrial Revolution. United Nations Industrial Development Organization. Vienna; 2019. 79 p.

[20] Hu H, Hong Y. Algal-bloom control by allelopathy of aquatic macrophytes—A review. Frontiers of Environmental Science & Engineering in China. 2008;2(4):421-438

[21] Macías FA, Galindo JLG, García-DíazMD,GalindoJCG.Allelopathic agents from aquatic ecosystems: potential biopesticides models. Phytochem Rev. 2008;7:155-178. DOI: 10.1007/s11101-007-9065-1

[22] Kurashov EA, Krylova JV, Mitrukova GG, Chernova AM. Low-molecular-weight metabolites of aquatic macrophytes growing on the territory of Russia and their role in hydroecosystems. Contemporary Problems of Ecology. 2014;7(4):433-448. DOI: 10.1134/S1995425514040064

[23] Burford MA, Gobler CJ, Hamilton DP, Visser PM, Lurling M, Codd GA. Solutions for managing cyanobacterial blooms: A scientific summary for policy makers. IOC/UNESCO, Paris; 2019. 17 p. (IOC/INF-1382)

[24] Zhu X, Dao G, Tao Y, Zhan X, Hu H. A review on control of harmful algal blooms by plant-derived allelochemicals. Journal of Hazardous Materials. 2021;401:123403. DOI:10.1016/j.jhazmat.2020.123403

[25] Mohamed ZA. Macrophytes-cyanobacteria allelopathic interactions and their implications for water resources management—A review. Limnologica - Ecology and Management of Inland Waters. 2017;63:122-132. DOI:10.1016/j.limno.2017.02.006

[26] Cheng F, Cheng ZH. Research progress on the use of plant allelopathy in agriculture and the physiological and ecological mechanisms of allelopathy. Front. Plant Sci. 2015;6:16. DOI:10.3389/fpls.2015.01020

[27] Saraf M, Pandya U, Thakkar A. Role of allelochemicals in plant growth promoting rhizobacteria for biocontrol of phytopathogens. Microbiol. Res. 2014;169:18-29. DOI: 10.1016/j.micres.2013.08.009

[28] Kong Ch. Allelopathic Research in China. Allelopathy Journal [Internet]. 2005. Available from: https://www.researchgate.net/publication/271885916_Allelopathy_in_China. [Accessed: 2020-11-23]

[29] Molisch H. Der Einfluss einer Pflanze auf die andere-Allelopathie. Fischer: Jena, Austria; 1937. 116 p

[30] Li Z-H, Wang Q, Ruan X, Pan C-D, Jiang D-A. Phenolics and Plant Allelopathy. Molecules. 2010;15(12):8933-8952. DOI:10.3390/molecules15128933

[31] Tokin BP. Über die mitogenetischen Strahlen und die Lisegangschen Ringe. Biol. Zentralblatt. 1930;50(11):641-671

[32] Tokin BP. Vegetable bactericides (phytoncides). Moscow: Medgiz; 1942. 108 p. [Bakteritsidy rastitel'nogo proiskhozhdeniya (fitontsidy)] (In Russ.)

[33] Gurevich FA. Relationship between the plants and freshwater animal embryos. Dokl. Akad. Nauk SSSR. 1948;59(3):569-572

[34] Gurevich FA. Phytoncides of aquatic and coastal plants, their role in hydrobiocenoses: dissertation abstract of Dr. Biol. Sciences: 030018 / Gurevich Faiva Abramovich; scientific consultant B. P. Tokin. Irkutsk; 1973. 29 p.

[35] Hasler AD, Jones E. Demonstration of the antagonistic action of large aquatic plants on algae and rotifers. Ecology. 1949;30:359-364. DOI:10.2307/1932616

[36] Kurashov EA, Mitrukova GG, Krylova JV. Interannual variability of low-molecular metabolite composition in Ceratophyllum demersum (Ceratophyllaceae) from a Floodplain lake with a changeable trophic status. Contemporary Problems of Ecology. 2018;11(2):179-194. DOI: 10.1134/S1995425518020063

[37] Van Donk E, van de Bund WJ. Impact of submerged macrophytes including charophytes on phyto- and zooplankton communities: Allelopathy versus other mechanisms. Aquat. Bot. 2002;72:261-274

[38] Seto M, Takamura N, Iwasa Y. Individual and combined suppressive effects of submerged and floating-leaved macrophytes on algal blooms. Journal of Theoretical Biology. 2013;319:122-133. DOI:10.1016/j.jtbi.2012.11.016

[39] Scheffer M, Hosper SH, Meijer M-L, Moss B, Jeppesen E. Alternative equilibria in shallow lakes. Trends Ecol. Evol. 1993;8:275-279

[40] Scheffer M. Alternative Attractors of Shallow Lakes. The Scientific World. 2001;1:254-263. DOI 10.1100/tsw.2001.62

[41] Scheffer M, Carpenter S, Foley JA, Folke C, Walker B. Catastrophic shifts in ecosystems. Nature. 2001;413:591-596

[42] Scheffer M, Szabo S, Gragnani A, van Nes EH, Rinaldi S, Kautsky N, Norberg J, Roijackers RMM, Franken RJM. Floating plant dominance as a stable state. Proc. Nat. Acad. Sci. 2003;100(7):4040-4045

[43] Jackson LJ. Macrophyte-Dominated and Turbid States of Shallow Lakes: Evidence from Alberta Lakes. Ecosystems. 2003; 6:213-223. DOI: 10.1007/s10021-002-0001-3

[44] Scheffer M. Ecology of Shallow Lakes. Springer US; 1998. 358 p.

[45] Mjelde M, Faafeng BA. Ceratophyllum demersum hampers phytoplankton development in some small Norwegian lakes over a wide range of phosphorus concentrations and geographical latitude. Freshwater Biol. 1997;37:355-365

[46] Van den Berg MS, Coops H, Meijer ML, Scheffer M, Simons J. Clear water associated with a dense Chara vegetation in the shallow and turbid lake Veluwemeer, The Netherlands. In: Jeppesen E, Sondergaard Ma, Sondergaard Mo and Christoffersen K, editors. The Structuring Role of Submerged Macrophytes in Lakes. New York: Springer; 1998. p. 339-352

[47] Blindow I, Hargeby A, Andersson G. Seasonal changes of mechanisms maintaining clear water in a shallow lake with abundant Chara vegetation. Aquat. Bot. 2002;72:315-334

[48] Hilt S, Gross EM. Can allelopathically active submerged macrophytes stabilize clear water states in shallow lakes? Basic Appl. Ecol. 2008;9:422-432

[49] Mulderij G, Mau B, van Donk E, Gross EM. Allelopathic activity of Stratiotes aloides on phytoplankton-towards identification of allelopathic substances. Hydrobiologia. 2007;584: 89-100. DOI:10.1007/s10750-007-0602-0

[50] Shao JH, Wu ZX, Yu GL, Peng X, Li RH. Allelopathic mechanism of pyrogallol to Microcystis aeruginosa PCC7806 (Cyanobacteria): from views of gene expression and antioxidant system. Chemosphere. 2009;75:924-928

[51] Hong Y, Hu H, Xie X, Sakoda A. Gramine-induced growth inhibition, oxidative damage and antioxidant responses in freshwater cyanobacterium Microcystis aeruginosa. Aquat. Toxicol. 2009;91(3):262-269

[52] Mohamed ZA, Al-Shehri AM. Differential responses of epiphytic and planktonic toxic cyanobacteria to allelopathic substances of the submerged macrophyte Stratiotes aloides. Int. Rev. Hydrobiol. 2010;95:224-234

[53] Zhu J, Liua B, Wang J, Gao Y, Wu Z. Study on the mechanism of allelopathic influence on cyanobacteria and chlorophytes by submerged macrophyte (Myriophyllum spicatum) and its secretion. Aquat. Toxicol. 2010;98 (2):196-203. DOI:10.1016/j.aquatox. 2010.02.011

[54] Nakai S, Yoshihara T, Yamada S, Hosomi M. The allelochemicals accounting for the allelopathic effects of Myriophyllum spicatum on the cyanobacterium Microcystis aeruginosa [Internet]. 2005. Available from: http://www.regional.org.au/au/allelopathy/2005/2/1/2244_nakais.htm#TopOfPage [Accessed: 2020-11-23]

[55] Nakai S, Zou G, Okuda T, Nishijima W, Hosomi M, Okada M. Polyphenols and fatty acids responsible for anti-cyanobacterial allelopathic effects of submerged macrophyte Myriophyllum spicatum. Water Sci. Technol. 2012;66:993-999

[56] Wang LX, Zhang L, Zhang YX, Jin CY, Lu CM, Wu GR. The inhibitory effect of Hydrilla verticillata culture water on Microcystic aeruginosa and its mechanism. J. Plant Physiol. Mol. Biol. 2006;32(6):672-678

[57] Wang HQ, Liang F, Qiao N, Dong JX, Zhang LY, Guo YF. Chemical composition of volatile oil from two emergent plants and their algae inhibition activity. Pol. J. Environ. Stud. 2014;6:2371-2374

[58] Wang H, Liang F, Zhang L. Composition and anti-cyanobacterial activity of essential oils from six different submerged macrophytes. Pol. J. Environ. Stud. 2015;24(1):333-338. DOI: https://doi.org/10.15244/pjoes/26383

[59] He CQ, Ye JX. Inhibitory effects of Acorus tatarinowii on algae growth. Acta Ecol. Sin. 1999;19(5):754-758

[60] Greca MD, Monaco P, Previtera L, Aliotta G, Pinto G, Pollio A. A llelochemical activity of phenylpropanes from Acorus gramineus. Phytochemistry. 1989;28(9):2319-2321 DOI:10.1016/s0031-9422(00)97975-5

[61] Foguel D, Chaloub RM. Effects of the alkaloid gramine on the light-harvesting, energy transfer, and growth of Anabaena sp. (PCC 7119). Plant Physiol. 1993;101:633-639

[62] Hong Y, Hu HY. Effects of the aquatic extracts of Arundo donax L. on the growth of freshwater algae. Allelopathy J. 2007;20:315-325

[63] Hong Y, Hu H-Y, Sakoda A, Sagehashi M. Isolation and characterization of antialgal allelochemicals from Arundo donax L. Allelopathy Journal. 2010;25(2):357-368

[64] Elakovich SD, Wooten JW. An examination of the phytotoxicity of the water shield Brasenia schreberi. J. Chem. Ecol. 1987;13(9):1935-1940. DOI:10.1007/bf01014676

[65] Nakai S, Inoue Y, Hosomi M, Murakami A. Growth inhibition of blue-green algae by allelopathic effects of macrophytes. Japanese Journal of Water Treatment Biology. 1997;33(4):215-222. DOI:10.2521/jswtb.33.215

[66] Nakai S, Inoue Y, Hosomi M. Growth inhibition of blue-green algae by allelopathic effects of macrophytes. Water Sci. Tech. 1999;39(8):47-53

[67] Zhou L, Chen G, Cui N, Pan Q, Song X, Zou G. Allelopathic effects on Microcystis aeruginosa and allelochemical identification in the cuture solutions of typical artificial floating-bed plants. Bull. Environ. Contam. Toxicol. 2019;102(1):115-121. DOI:10.1007/s00128-018-2486-2

[68] Wium-Andersen S. Allelopathy among aquatic plants. Archiv für Hydrobiologie, Beiheft Ergebnisse der Limnologie. 1987;27:167-172

[69] Kogan SI, Chinnova GA. Relations between Ceratophyllum demersum L. and some blue-green algae. Hydrobiol. J. 1972;8:14-19

[70] Jasser I. Influence of Ceratophyllum demersum on phytoplankton community in experimental conditions. Verh. Internat. Verein. Limnol. 1994;25: 2291-2295

[71] Jasser I. The influence of macrophytes on a phytoplankton community in experimental condition. Hydrobiologia. 1995;306:21-32. DOI:10.1007/BF00007855

[72] Korner S, Nicklisch A. Allelopathic growth inhibition of selected phytoplankton species by submerged macrophytes. Journal of Phycology. 2002;38(5):862-871. DOI:10.1046/j.1529-8817.2002.t01-1-02001.x

[73] Wium-Andersen S, Anthoni U, Houen G. Elemental sulphur, a possible allelopathic compound from Ceratophyllum demersum. Phytochemistry. 1983;22:2613

[74] Gross EM, Erhard D, Iványi E. Allelopathic activity of Ceratophyllum demersum L. and *Najas marina ssp. intermedia* (Wolfgang) Casper. Hydrobiologia. 2003;506-509(1-3): 583-589. DOI:10.1023/b:hydr.0000008539.32622.91

[75] Qiming X, Haidong C, Huixian Z, Daqiang Y. Allelopathic activity of volatile substance from submerged macrophytes on Microcystin aeruginosa. Acta Ecologica Sinica. 2006;26(11):3549-3554

[76] Ghobrial MG, Nassr HS, Kamil AW. Bioactivity effect of two macrophyte extracts on growth performance of two bloom-forming cyanophytes. The Egyptian Journal of Aquatic Research. 2015;41(1):69-81. DOI:10.1016/j.ejar.2015.01.001

[77] Amorim CA, de Moura-Falcão RH, Valença CR, de Souza VR, Moura AN. Allelopathic effects of the aquatic macrophyte Ceratophyllum demersum L. on phytoplankton species: contrasting effects between cyanobacteria and chlorophytes. Acta Limnol. Bras. 2019;31. DOI: 10.1590/s2179-975x1419

[78] Tazart Z, Douma M, Caldeira AT, Tebaa L, Mouhri K, Loudiki M. Highlighting of the antialgal activity of organic extracts of Moroccan macrophytes: potential use in cyanobacteria blooms control. Environ Sci Pollut Res. 2020;27:19630-19637. DOI:10.1007/s11356-020-08440-w

[79] Berger J, Schagerl M. Allelopathic activity of Characeae. Biologia. 2004;59:9-15

[80] Berger J, Schagerl M. Allelopathic acitivity of Chara aspera. Hydrobiologia. 2003;501:109-115

[81] Złoch I, Śliwińska-Wilczewska S, Kucharska M, Kozłowska W. Allelopathic effects of Chara species (*C. aspera*, *C. baltica*, and C. canescens) on the bloom-forming picocyanobacterium Synechococcus sp. Environ Sci Pollut Res. 2018;25:36403-36411. DOI:10.1007/s11356-018-3579-5

[82] Pakdel FM, Sim L, Beardall J, Davis J. Allelopathic inhibition of microalgae by the freshwater stonewort, Chara australis, and a submerged angiosperm, Potamogeton crispus. Aquatic Botany. 2013;110:24-30. DOI:10.1016/j.aquabot.2013.04.005

[83] Rojo C, Segura M, Rodrigo MA. The allelopathic capacity of submerged macrophytes shapes the microalgal assemblages from a recently restored coastal wetland. Ecol. Eng. 2013;58:149-155. DOI:10.1016/j.ecoleng.2013.06.019

[84] Wium-Andersen S, Anthoni U, Christophersen C, Houen G. Allelopathic effects on phytoplankton by substances isolated from aquatic macrophytes (Charales). Oikos. 1982;39:187-190

[85] Horecka M. The significant role of Chara hispida grown in water regions of a gravel pit lake. Senec. Arch. Protistenkd. 1991;139:275-278

[86] Crawford SA. Farm pond restoration using Chara vulgaris vegetation. Hydrobiologia. 1979;62(1):17-31. DOI:10.1007/bf00012559

[87] Zhang T, He M, Wu A, Nie L. Allelopathic effects of submerged macrophyte Chara vulgaris on toxic Microcystis aeruginosa. Allelopath. J. 2009;23(2):391-401

[88] Yang SY, Sun WH. Isolation and identification of antialgal compounds from root system of water hyacinth. Acta Photophysiol. Sin. 1992;18(4):399-402

[89] Shanab SMM, Shalaby EA, Lightfoot DA, El-Shemy HA. Allelopathic effects of water hyacinth [Eichhornia crassipes]. PLoS ONE. 2010;5(10):e13200. DOI:10.1371/journal.pone.0013200

[90] Liu R, Ran X, Bai F, Xu J, Yang S, Shi J, Wu Z. Use of chlorophyll a fluorescence to elucidate the toxicity target of N-phenyl-2-naphthylamine on photosynthetic system of Cylindrospermopsis raciborskii (Cyanobacteria). Phycologia. 2015;54:12-19

[91] Pei Y, Liu L, Hilt S, Xu R, Wang B, Li C, Chang X. Root exudated algicide of Eichhornia crassipes enhances allelopathic effects of cyanobacteria Microcystis aeruginosa on green algae. Hydrobiologia. 2018;823:67-77. DOI:10.1007/s10750-018-3696-7

[92] Proctor VW. Some controlling factors in the distribution of *Haematococcus pluvialis*. Ecology. 1957;38:457-462

[93] Van Aller RT, Pessoney GF, Rogers VA, Watkins EJ, Leggett HG. Oxygenated fatty acids: A class of allelochemicals from aquatic plants. ACS Symp. Ser. 1985;268:387-400. DOI: 10.1021/bk-1985-0268.ch026

[94] Erhard D, Gross EM. Allelopathic activity of Elodea canadensis and Elodea nuttallii against epiphytes and phytoplankton. Aquatic Botany. 2006;85(3):203-211. DOI:10.1016/j.aquabot.2006.04.002

[95] Wang HQ, Zhu HJ, Zhang LY, Xue WJ, Yuan B. Identification of antialgal compounds from the aquatic plant Elodea nuttallii. Allelopathy J. 2014;34:207-213

[96] Gao H, Song Y, Lv C, Chen X, Yu H, Peng J, Wang M. The possible allelopathic effect of Hydrilla verticillata on phytoplankton in nutrient-rich water. Environmental Earth Sciences. 2015;73(9):5141-5151. DOI:10.1007/s12665-015-4316-8

[97] Cheng W, Xuexiu C, Hongjuan D, Difu L, Junyan L. Allelopathic inhibitory effect of Myriophyllum aquaticum (Vell.) Verdc. on Microcystis aeruginosa and its physiological mechanism. Acta Ecologica Sinica. 2008;28(6):2595-2603

[98] Saito K, Matsumoto M, Sekine T, Murakoshi I, Morisaki N,

Iwasaki S. Inhibitory substances form Myriophyllum brasiliense on growth of bluegreen algae. J. Nat. Prod. 1989;52(6):1221-1226. DOI:10.1021/np50066a004

[99] Bi YL, Wu SM, Zhou SN, Wu SH, Xu SJ. Allelopathic effects and allelochemicals of Myriophyllum elatinoides on Microcystis aeruginosa and *Selenastrum capricornutum*. Environ. Sci. 2019;40:2265-2270. DOI: 10.13227/j.hjkx.201810064

[100] Planas D, Sarhan F, Dube L, Godmaire H. Ecological significance of phenolic compounds of Myriophyllum spicatum. Verh. Internat. Verein. Limnol. 1981;21:1492-1496

[101] Gross EM, Sütfeld R. Polyphenols with algicidal activity in the submerged macrophytes Myriophyllum spicatum L. Acta Hortic. 1994;381:710-716

[102] Nakai S, Yamada S, Hosomi M. Anticyanobacterial fatty acids released from Myriophyllum spicatum. Hydrobiologia. 2005;543:71-78

[103] Gross EM, Meyer H, Schilling G. Release and ecological impact of algicidal hydrolysable polyphenols in Myriophyllum spicatum. Phytochemistry. 1996;41(1):133-138

[104] Gross EM. Allelopathy in benthic and littoral areas: case studies on allelochemicals from benthiccyano-bacteria and submerged macrophytes. In: Inderjit Dakshini KMM, Foy CL, editors. Principles and Practices in Plant Ecology: Allelochemical Interactions. CRC Press, Boca Raton, FL; 1999. p. 179-199

[105] Aliotta G, Molinaro A, Monaco P, Pinto G, Previtera L. Three biologically active phenylpropanoid glucosides from Myriophyllum verticillatum. Phytochemistr. 1992;31(1):109-111

[106] Pollio A, Pinto G, Ligrone R, Aliotta G. Effects of the potential allelochemical a-asarone on growth, physiology and ultrastructure of two unicellular green algae. J. Appl. Phycol. 1993;5:395-403

[107] Tazart Z, Caldeira AT, Douma M, Salvador C, Loudiki M. Inhibitory effect and mechanism of three macrophytes extract on Microcystis aeruginosa growth and physiology. Water and Environment Journal. 2020. DOI:10.1111/wej.12653

[108] Li FM, Hu HY. Isolation and characterization of a novel antialgal allelochemical from Phragmites communis. Applied and Environmental Microbiology. 2005;71(11):6545-6553. DOI:10.1128/aem.71.11.6545-6553.2005

[109] He L, Meng F-L, Meng R, Huang C-H, Li Y-W, Xi B-D, Shu J-M. In Situ Enclosure Experiment on Nelumbo nucifera for Eutrophication Control in Baiyang dian Lake. Wetland Science. 2013;11(2):282-285

[110] Balanda OV, Sakevich AI. Alkaloids of Nuphar lutea (L.) Smith. and their Influence on the Vital Activity of Cyanobacteria and Algae. Hydrobiological Journal. 2004;40(6):108-120. DOI: 10.1615/HydrobJ.v40.i6.100

[111] Ahluwalia AS, Ghawana VK. Allelopathic interactions among cyanobacteria and rice and their potential role in biotechnology. In: Subramanian G, Kaushik BD, Venkataraman GS, editors. Cyano-bacterial Biotechnology. India: Oxford and IBH Publishing;1998. p. 403-408

[112] Nakai S, Zhou S, Hosomi M, Tominaga M. Allelopathic growth inhibition of cyanobacteria by reed. Allelopathy J. 2006;18(2):277-286

[113] Aliotta G, Monaco P, Pinto G, Pollio A, Previtera L. Potential allelochemicals from Pistia stratiotes L. J. Chem. Ecol. 1991;17(11):2223-2234. DOI:10.1007/bf00988003

[114] Wu X, Wu H, Chen J, Ye J. Effects of allelochemical extracted from water lettuce (Pistia stratiotes Linn.) on the growth, microcystin production and release of Microcystis aeruginosa. Environmental Science and Pollution Research. 2013;20(11):8192-8201. DOI:10.1007/s11356-013-1783-x

[115] Wu X, Wu H, Ye J, Zhong B. Study on the release routes of allelochemicals from Pistia stratiotes Linn., and its anti-cyanobacteria mechanisms on Microcystis aeruginosa. Environmental Science and Pollution Research. 2015;22(23):18994-19001. DOI:10.1007/s11356-015-5104-4

[116] Zhou Y, Zhou X, Han R, Xu X, Wang G, Liu X, Feng D. Reproduction capacity of Potamogeton crispus fragments and its role in water purification and algae inhibition in eutrophic lakes. Science of The Total Environment. 2017;580:1421-1428. DOI:10.1016/j.scitotenv.2016.12.108

[117] Kang LJ, Xu H, Zou W, Zhu GW, Zhu MY, Ji PF, Chen J. Influence of Potamogeton crispus on Lake Water Environment and Phytoplankton Community Structure. Environmental Science. 2020;41(9):4053-4061. DOI: 10.13227/j.hjkx.202002006

[118] Zhang S, Cheng S, Wang HQ, He F, Wu ZB. Allelopathic interactions between the Potamogeton spp and toxic cyanobacteria (Microcystis aeruginosa). Allelopathy Journal. 2009;23(2):379-390

[119] Zhang S, Sun P, Ge F, Wu Z. Different Sensitivities of *Selenastrum capricornutum* and Toxic Strain Microcystis aeruginosa to Exudates from Two Potamogeton Species. Pol. J. Environ. Stud. 2011;20(5):1359-1366

[120] Zhang W, Shen H, Zhang J, Yu J, Xie P, Chen J. Physiological differences between free-floating and periphytic filamentous algae, and specific submerged macrophytes induce proliferation of filamentous algae: A novel implication for lake restoration. Chemosphere. 2020;239: 124702. DOI:10.1016/j.chemosphere.2019.124702

[121] Dehui C, Yongding L, Lirong S. The allelopathy of macrophyte Potamogeton pectinatus L on chlorophyta (*Scenedesmus obliquus*) and cyanobacteria (Microcystis aeruginosa) and calculation of allelopathic parameter. Acta Hydrobiologica Sinica. 2004;28(2):163-168

[122] Zhang X, Lu X, Wang L. Allelopathic effect of Ruppia maritima on *Chlorella vulgaris* and Microcystis aeruginosa. China Environmental Sciencece. 2019; 39(4): 1589-1595

[123] Greca MD, Fiorentino A, Isidori M, Monaco P, Zarrelli A. Antialgal ent-labdane diterpenes from Ruppia maritime. Phytochemistry. 2000;55: 909-913

[124] Dai SG, Zhao F, Jin ZH, Zhuang YY, Yuan YC. Allelopathic effect of plant's extracts on algae and the isolation and identification of phytotoxins. Environ. Chem. 1997;16(3):268-271

[125] Aliotta G, Greca MD, Monaco P, Pinto G, Pollio A, Previtera L. In vitro algal growth inhibition by phytotoxins of Typha latifolia L. J. Chem. Ecol. 1990;16(9):2637-2646. DOI: 10.1007/BF00988075

[126] Greca MD, Mangoni L, Molinaro A, Monaco P, Previtera L. (20S)-4α-methyl-24-methylenecholest-7-en-3β-ol, an allelopathic sterol from Typha latifolia. Phytochemistry. 1990;29:1797-1798. DOI:10.1016/0031-9422(90)85019-c

[127] Xian QM, Chen HD, Liu HL, Zou HX, Yin DQ. Isolation and identification of antialgal compounds from the leaves of Vallisneria spiralis L. by activity-guided fractionation. Environ. Sci. Pollut. Res. 2006;13(4):233-237

[128] Li B, Yin Y, Kang L, Feng L, Liu Y, Du Z, Tian Y, Zhang L. A review: Application of allelochemicals in water ecological restoration – algal inhibition. Chemosphere. 2020;128869. DOI:10.1016/j.chemosphere.2020.128869

[129] Lombardo P, Mjelde M, Källqvist T, Brettum P. Seasonal and scale-dependent variability in nutrient- and allelopathy-mediated macrophyte–phytoplankton interactions. Knowledge and Management of Aquatic Ecosystems. 2013;409:10. DOI:10.1051/kmae/2013055

[130] Mulderij G, Mooij WM, Smolders AJP, Van Donk E. Allelopathic inhibition of phytoplankton by exudates from Stratiotes aloides. Aquatic Botany. 2005;82:284-296. DOI:10.1016/j.aquabot.2005.04.001

[131] Elakovich, SD. Allelopathic aquatic plants for aquatic weed management. Biol Plant. 1989;31:479. DOI:10.1007/BF02876221

[132] Elakovich SD, Wooten JW. Allelopathic potential of Nuphar lutea (L.) Sibth. & Sm. (Nymphaeaceae). J Chem Ecol. 1991;17:707-714. DOI:10.1007/BF00994194

[133] Dekker J, Meggitt WF. Interference between velvetleaf (*Abutilon theophrasti* Medic.) and soybean (*Glycine max* (L.) Merr.). I. Growth. Weed Res. 1983;23:91-101

[134] Ball AS, Williams M, Vincent D. Algal growth control by a barley straw extract. Bioresource Technol. 2001;77:177-181

[135] Everall NC, Lees DR. The identification and significance of chemical released from decomposing barley straw during reservoir algal control. Water Res. 1997;31:614-620

[136] Ferrier MD, Butler BR Sr, Terlizzi DE. The effects of barley straw (*Hordeum vulgare*) on the growth of freshwater algae. Bioresource Technol. 2005;96:1788-1795

[137] Xiao X, Huang H, Ge Z, Rounge TB, Shi J, Xu X, Li R, Chen Y. A pair of chiral flavonolignans as novel anti-cyanobacterial allelochemicals derived from barley straw (*Hordeum vulgare*): characterization and comparison of their anti-cyanobacterial activities. Environ. Microbiol. 2014;16 (5):1238-1251. DOI:10.1111/1462-2920.12226

[138] Kurashov EA, Fedorova EV, Krylova JV, Mitrukova GG. Assessment of the potential biological activity of low molecular weight metabolites of freshwater macrophytes with QSAR. Scientifica. 2016; Article ID 1205680. DOI:10.1155/2016/1205680

[139] Kurashov EA, Fedorova EV, Krylova JV. Using the QSAR method to identify the most promising allelochemics against cyanobacteria. [Ispol'zovaniye metoda QSAR dlya vyyavleniya naiboleye perspektivnykh allelokhemikov v otnoshenii tsianobakteriy]. Russian Journal of Applied Ecology. 2018;4(16):56-61 (In Russ.)

[140] Kurashov E, Kapustina L, Krylova J, Mitrukova G. The use of fluorescence microscopy to assess the suppression of the development of cyanobacteria under the influence of allelochemicals of aquatic macrophytes. In: Natalia Grigoryeva, editor. Fluorescence Methods for Investigation of Living Cells and Microorganisms. IntechOpen; 2020. 28 p. DOI: 10.5772/intechopen.92800

[141] Ni LX, Jie XT, Wang PF, Li SY, Wang GX, Li YP, Li Y Acharya K. Effect of linoleic acid sustained-release microspheres on Microcystis aeruginosa antioxidant enzymes activity and microcystins production and release.

Chemosphere. 2015;121:110-116. DOI: 10.1016/j.chemosphere.2014.11.056

[142] Ni LX, Jie XT, Wang PF, Li SY, Hu SZ, Li YP, Li Y Acharya K. Characterization of unsaturated fatty acid sustained-release microspheres for long-term algal inhibition. Chemosphere. 2015b;120:383-390. DOI:10.1016/j.chemosphere.2014.07.098

[143] Ni LX, Acharya K, Ren GX, Li SY, Li YP, Li Y. Preparation and characterization of anti-algal sustained-release granules and their inhibitory effects on algae. Chemosphere. 2013;91:608-615. DOI:10.1016/j.chemosphere.2012.12.064

[144] Huang HM, Xiao X, Lin F, Grossart HP, Nie ZY, Sun LJ, Xu C, Shi JY. Continuous-release beads of natural allelochemicals for the long-term control of cyanobacterial growth: preparation, release dynamics and inhibitory effects. Water Res. 2016;95:113-123. DOI:10.1016/j.watres.2016.02.058

[145] Gao Y, Liu B, Ge FJ, He Y, Lu Z, Zhou Q, Zhang YY, Wu ZB Joint effects of allelochemical nonanoic acid, N-phenyl-1-naphtylamine and caffeic acid on the growth of Microcystis aeruginosa. Allelopathy Journal. 2015; 35(2):249-258

[146] Zuo S, Zhou S, Ye L, Ma S. Synergistic and antagonistic interactions among five allelochemicals with antialgal effects on bloom-forming Microcystis aeruginosa. Ecological Engineering. 2016;97:486-492. DOI:10.1016/j.ecoleng.2016.10.013

[147] Wang HQ, Zhang LY. Allelopathic activity of ethyl acetate extracts from typical emergent plants against Microcystis aeruginosa Kütz. Bangladesh J. Bot. 2017;46: 025-1029

[148] Nakai S, Inoue Y, Hosomi M, Murakami *A. myriophyllum* spicatum-released allelopathic polyphenols inhibiting growth of blue-green algae Microcystis aeruginosa. Water Res. 2000;34(11):3026-3032. DOI:10.1016/S0043-1354(00)00039-7

[149] Nakai S, Inoue Y, Hosomi M. Algal growth inhibition effects and inducement modes by plant-producing phenols. Water Res. 2001;35:1855-1859. DOI:10.1016/S0043-1354(00)00444-9

[150] Gao Y-N, Liu B-Y, Xu D, Zhou Q-H, Hu C-Y, Ge F-J, Zhang L-P, Wu Z-B. Phenolic compounds exuded from two submerged freshwater macrophytes and their allelopathic effects on Microcystis aeruginosa. Polish J. Environ. Stud. 2011;20:1153-1159

[151] Greca MD, Lanzetta R, Mangoni L, Monaco P, Previtera L. A bioactive benzoindenone from Eichhornia crassipes. Solms. Bioorg. Med. Chem. Lett. 1991;1:599-600

[152] Greca MD, Lanzetta R, Molinaro A, Monaco P, Previtera L. Phenalene metabolites from Eichhornia crassipes. Bioorg. Med. Chem. Lett. 1992; 2:311-314

[153] Greca MD, Fiorentino A, Monaco P, Pinto G, Pollio A, Previtera L. Action of antialgal compounds from Juncus effuses L. on *Selenastrum capricornutum*. J. Chem. Ecol. 1996; 22(3):587-603

[154] Greca MD, Fiorentino A, Monaco P, Pinto G, Previtera L, Zarrelli A. Synthesis and antialgal activity of dihydrophenanthrenes and phenanthrenes II: Mimics of naturally occurring compounds in *Juncus effusus*. J. Chem. Ecol. 2001;27(2):257-271

[155] Sun WH, Yu SW, Yang SY, Zhao BW, Yu ZW, Wu HL, Huang SY, Tang CS. Allelochemicals from root exudates of water hyacinth (Eichhornis crassipes). Acta Photophysiol. Sin. 1993;19(1):92-96

[156] Sutfeld R, Petereit F, Nahrstedt A. Resorcinol in exudates of Nuphar lutea. J. Chem. Ecol. 1996;22:2221-2231

[157] Sutfeld R. Polymerization of resorcinol by a cryptophycean exo-enzyme. Phytochemistry. 1998;49: 451-459

[158] Gallardo MT, Martin BB, Martin DF. Inhibition of water ferm (Salviania minima) by cattail (*Typha domingensis*) extracts and by 2-chlorophenol and salicylaldehyde. J.Chem. Ecol. 1998;24:1483-1490

[159] Della GM, Ferrara M, Fiorentino A, Monaco P, Previtera L. Antialgal compounds from *Zantedeschia aethiopica*. Phytochemistry. 1998;49 (5):1299-1304

[160] Choe S, Jung IH. Growth inhibition of freshwater algae by ester compounds released from rotted plants. J. Ind. Eng. Chem. 2002;8(4):297-304

[161] Nakai S, Hosomi M. Allelopathic inhibitory effects of polyphenols released by Myriophyllum spicatum on algal growth. Allelopath. J. 2002;10(2):123-131

[162] Wu ZB, Deng P, Wu XH, Luo S, Gao YN. Allelopathic effects of the submerged macrophyte Potamogeton malaianus on *Scenedesmus obliquus*. Hydrobiologia. 2007;592: 65-474

[163] Liu H, He L, Gao J, Ma Y, Zhang X, Peng H, Chen J. Chemical constituents from the aquatic weed Pistia stratiotes. Chem. Nat. Compd. 2008;44(2):236-238

[164] Liu G, Zhou C, Sun L, Zhu W, Jiang H, Wang H, An S. Effects of Eichhornia crassipes allelochemicals on the growth of two mono- and co-cultured algae Microcystis aeruginosa and Scenedesmus obliquu. Acta Scientiae Circumstantiae. 2011;31(10):2303-2311

[165] Huang H, Xiao X, Ghadouani A, Wu J, Nie Z, Peng C, Shi J. Effects of Natural Flavonoids on Photosynthetic Activity and Cell Integrity in Microcystis aeruginosa. Toxins. 2015;7(1):66-80. DOI:10.3390/toxins7010066

[166] Dong J, Chang M, Li C, Dai D, Gao Y. Allelopathic effects and potential active substances of Ceratophyllum demersum L. on *Chlorella vulgaris* Beij. Aquatic Ecol. 2019;53(4):651-663

[167] Kurashov EA, Krylova JV, Bataeva YuV, Rusanov AG, Sukhenko LT. Algicide for suppressing the development of cyanobacteria and green algae based on metabolites-allelochemicals of aquatic plants. Patent for invention RU 2709308 C1, 17.12.2019. Application No. 2019104959 dated 02.21.2019. [Internet]. 2019. Available from: https://patents.s3.yandex.net/RU2709308C1_20191217.pdf [Accessed: 2020-11-23]

[168] Gross EM, Hilt S, Lombardo P, Mulderij G. Searching for allelopathic effects of submerged macrophytes on phytoplankton-state of the art and open questions. Hydrobiology. 2007;584:77-88

[169] Hilt S, Ghobrial M, Gross EM. In situ allelopathic potential of Myriophyllum verticillatum (Haloragaceae) against selected phytoplankton species. J. Phycol. 2006;42:1189-1198

[170] Vanderstukken M, Mazzeo N, Van Colen W, Declerck SAJ, Muylaert K. Biological control of phytoplankton by the subtropical submerged macrophytes Egeria densa and *Potamogeton illinoensis*: a mesocosm study. Freshwater Biology. 2011;56(9):1837-1849. DOI:10.1111/j.1365-2427.2011.02624.x

[171] Švanys A, Paškauskas R, Hilt S. Effects of the allelopathically active macrophyte Myriophyllum spicatum on a natural phytoplankton community: a mesocosm study. Hydrobiologia. 2013;737(1):57-66. DOI:10.1007/s10750-013-1782-4

[172] Patova EN. Bloom-forming Cyano-prokaryotes in Kharbeyskie Lakes of

Bolshezemelskaya Tundra. Journal of Siberian Federal University. Biology. 2014;7(3):282-290

[173] Beljakova RN. Bloom forming cyanoprokaryota from water bodies of North-Western Russia. Novitates Systematicae Plantarum Non Vascularium. 2005;39:12-36. (In Russ.)

[174] Skulberg OM, Underdal B, Utkilen H. Toxic waterblooms with cyanophytes in Norway — current knowledge. Algological Studies. 1994;75:279-289

[175] Willis RJ. The History of Allelopathy. Springer, Dordrecht; 2007. 316 p. DOI: 10.1007/978-1-4020-4093-1

[176] Bauer N, Grossart HP, Hilt S. Effects of bacterial communities on the sensitivity of Stephanodiscus minutulus and Desmodesmus armatus to tannic acid. Aquat. Microb. Ecol. 2010;59: 295-306

Section 3

Zooplankton

Chapter 6

Food Webs

Laura Mondragón-Camarillo and Salvador Rodríguez Zaragoza

Abstract

Ciliates are important elements of the trophic networks of aquatic and terrestrial environments, they can be primary producers (myxotrophs), consumers of bacteria, algae, flagellates, even other ciliates and can serve as food for metazoans, for all the above they are the link between different levels of food webs. The structure of the ciliates varies according to the seasons of the year and depending on the trophic conditions of the aquatic systems. Ciliated communities have modifications and adaptations in response to environmental perturbations. The objective of this chapter is to describe the importance of different trophic groups of ciliates in different ecosystems, including anthropogenic perturbations and their impact on trophic webs.

Keywords: ciliates, trophic groups, food webs, perturbations

1. Introduction

Trophic relationships between organisms are the mechanisms responsible for most of energy and nutrient transfers; they allow the functioning of the ecosystem. These relationships, known as food webs, caught the attention of naturalists before the concepts of evolution and ecology were about to be determined.

Initially, the diet of a species and its skills to obtain it were recognized as the leading factors for the prevalence of the fittest. Additionally, it is one of the main forces leading to evolution of that species in the long term [1]. Furthermore, competition for food became one of the favorite hypotheses to explain species exclusion; it states that when two species seemed to feed on the same resources, the best suited ultimately leads its competitor to extinction in the long term [2]. This idea has been around for many years and has not been completely discarded or proved [1].

Examining phototrophs, also known as primary producers, is the dominant starting point to analyze food webs. They use the incoming sun's energy and inorganic nutrients to generate their biomass. This is the most important mechanism, as it initiates the cycling of nutrients and energy flux in aquatic food webs. There is primary productivity involving chemolithotrophs dominating in places devoid of sun's light [3]. These places were mostly known to be, until recent times, around underwater volcanoes more than 1000 meters deep [3, 4].

Primary production is at the base of all consumers concurring in the environment. However, macroscopic food webs tend to be very short, with few levels of consumers because these organisms dissipate matter and energy efficiently [5]. All metazoans invest their energy looking for food, ingesting it, digesting, repairing themselves, mating, and reproducing. These activities make multicellular organisms to get around 10% of biomass fixation efficiency. Thus, 1,000 kgs of the primary producer will be needed to produce 100 kgs of herbivorous animals, only 10 kgs of small carnivores, 1 kgs of medium-sized carnivores, and only 0.1 kgs of top carnivores, following a pattern known as pyramid of productivity [5]. Adding a predatory species at any level would destabilize the food web, as this will consume higher amounts of biomass [6]. Energy dissipation is even larger, meaning that the entropy produced during the functioning of the food web is very high. However, only 1% of the incoming sunlight is used for primary production, stressing the importance of the environmental factors limiting biomass productivity to sustain food webs.

Primary productivity varies along seasons. When it reaches its peak, productivity is controlled by the top predator's consumption (top-down), and when it reaches its lowest level, productivity is controlled by phototrophs (bottom-up). There are places that are permanently bottom-up controlled such as the deep ocean communities depending on the "organic matter rain" from dead organisms living in the photic zone in places near the equator are almost always top-down controlled, where productivity may be at its peak for most of the year. All other places experience top-down/bottom-up controls alternatively, depending on the productivity seasons.

Unicellular algae lead primary productivity in marine environments, sustaining the great diversity of organisms, especially in places receiving nutrient inputs from lands. Heterotrophic unicellular organisms forage on algae and both phototrophs (phytoplankton) and heterotrophs (zooplankton) conform to the plankton. However, unicellular organisms span in sizes less than 1 μm to hundreds of micrometers, and the species' diversity of plankton, including microbial eukaryotes and bacteria, ranges in the order of thousands. Species of microorganisms are much more numerous than the metazoans. With such a great diversity of microorganisms, it become apparent that the microbial food webs may function differently from the macroscopic food webs.

It was believed that food webs would get destabilized if the number of species increases at any level above the primary producers. However, microbial food web seemed to get more stability with the increasing number of species, contradicting what was observed in macroscopic food webs [7]. Thus, the higher number of species of bacterial and microbial eukaryotes in aquatic food webs seemed to contradict that assumption; this phenomenon was named as "The paradox of microbial loop." It was paradoxical that productivity and efficiency of nutrients and energy transformation is increased by adding more species, promoting the stabilization of the food web [8].

It's been a long road since the recognition of the "paradox of the microbial loop" in the aquatic food webs. Nowadays, it is referred only as the "microbial loop," after being integrated into the food web conceptualization in both terrestrial and aquatic environments [7].

The complexity of microbial food webs needs to be approached from the analysis of different functioning capacities and nutritional needs of the participating species. It has been normal to assign very general feeding habits to protists and metazoans, like bacterivores for example. This nomenclature implicates that a single species of protist can feed on any one or indistinctly on all the thousands of bacteria species. However, observation of feeding habits has revealed that protists and metazoans prefer feeding on specific kind of bacteria while avoiding other species. Pigmented bacteria [9], for example, has fewer predators than non-pigmented ones. On the other hand, there are several species of protists, mostly amoebae, small flagellates and *Colpoda steinii* that feed on pigmented ones [10].

One explanation for pigmented bacteria to have fewer predators relied on the toxicity or poisonous effect of those pigments for many protists, pointing out the importance of the biochemical warfare that bacteria must synthetize to defend themselves. However, chemicals used for evading enemies attract other ones looking for those same compounds, putting bacteria in a situation where there is no way out for bacterial preys. Indeed, there is no way out of being preyed upon, as every living being has predators, or at least other species which may feed on them or use them as a resource.

Is there a single factor determining the feeding preferences? The short answer is "No." Remember that "bacterivorous" or "algivorous" are labels used to recognize the kind of food that protists and metazoans may prefer to feed on, and it involves many species. From the beginning, this was a non-exclusive way to label the category of food that may be used to group the highest quantity of species to simplify and conceptualize the food webs. Furthermore, during the first half of the XX century [11], there were many very interesting studies trying to determine the "diets" of several species of protists [11, 12], with the aim of designing a chemically defined culture media, as is the case of several recipes for culturing *Tetrahymena pyriformis, Glaucoma sp,* or *Paramecium* sp., culminating with 3 books edited by Lewandowsky and Hutner (1979), approaching the field of protists' biochemistry (at that time it was biochemistry of protozoa).

Designing a culture media for protists or bacteria was a major task, as numerous factors about their nutritional needs were unknown (and remain unknown). These attempts to cultivate bacteria and protists lead to one important conclusion: different species cannot synthesize one or several molecules needed for their metabolism and have to take those molecules, as such, from their ingested food [12] or from other microorganisms that live within the biofilm (such as the case of NAD+ **, which the bacteria has to consume from other species of bacteria for both of them to grow). Microbial biologists named this phenomenon as "auxotroph" [13]. In this way, the molecule(s) a bacterial species is auxotroph for must be added to the culture media, to keep a culture of such species [14]. The kind of molecules, their diversity, and their macro- and micronutrient composition form a universe comparable to the one containing the species' diversity on the planet.

2. Phagotrophic protists

Ecological relevant functions have been recognized in prokaryotes and microbial eukaryotes. Bacteria have been cataloged as nitrogen fixing, denitrifiers, metanogens, methanotrophs, phosphorous mobilizers, metal mobilizers, phototrophs, and chemolithotrophs as the main recognized functions in the ecosystem. On the protists' side, several trophic groups have been recognized as phototrophs and phagotrophs. The first group is strictly divided between the phototrophs and mixotrophic ones, while the second one may be divided in bacterivorous (including cyanobacteria), frugivorous (feeding on hypha and or yeasts cells), algivorous, protist consumers (raptorial protists), and metazoan predators. Parasitic bacteria, pathogenic bacteria, and microbial eukaryotes have been largely studied from the medical point of view. However, recently, they have been studied from the ecological perspective (their impact on the predator–prey relationships, the "health" of species populations protected for conservation, and their effect on the nutrients distribution along food webs [13].

Phagotrophic protists may ingest very different kinds of particles and present the capacity to eject the ones they cannot digest, or even reject particles previously ingested [15]. Even if the water current would transport a good mixture of different bacterial species, phagotrophs may choose which particles ingest and eject the debris from their digestion together with the non-digestible microorganisms. This means that protists may show preference for the kind of food they most likely can digest (recognizing their preys by their quorum sensing signals), and, like bigger organisms, they may need a variety of food sources to get the nutrients they need [15].

A close examination of the different trophic groups allows to re-mark the unicellular phototrophs as the most productive in terms of biomass production since there is no synthesis of support or conductive structures, and, because of that, they are the base of the aquatic food webs.

The phagotrophic protists have been recognized for being the main consumers along microbial trophic networks in aquatic systems conforming a major proportion of the microbial biomass in these systems [16, 17]. These predators are also responsible for much of the recycling flow of nitrogen and phosphorus in the aquatic systems [18].

Particularly the ciliates are key elements of aquatic food webs they have several functions, they can be primary producers, predators, they serve as food for metazoans including free-living stages of metazoan parasites; there are many aquatic habitats without macro-organisms, but none without bacteria and at least few protist species [19].

2.1 Mixotrophic ciliates

One of the most interesting groups of protists are the mixotrophic ones. Some of them may correspond to the old morphological groups of ciliates, flagellates, and ameboebas. Mixotrophy is defined as the ability to combine phagotrophy and phototrophy in a single cell [20]. This group can be divided into constitutive mixotrophs, meaning they have the innate ability to photosynthesize, and the facultative or non-constitutive mixotrophs. These organisms may sequester the plastids after consuming their phototrophic preys or by harboring photosynthetic endosymbionts [20, 21]. Around 23% of planktonic ciliates species (marine and freshwater combined) perform acquired phototrophy, and this ability is present in at least 8 main ciliated taxa: Heterotrichea, Hypotrichia, Oligotrichida, Stichotrichida, Litostomatea, Prostomatea, Peniculia, and Peritrichia. Phototrophy is usually acquired from algae endosymbionts in 7 of these 8 ciliated taxa. Contrastingly, Oligotrichida usually obtains this ability by plastid sequestering [22].

The structures of the mixotrophic ciliates community varies through seasons, depending on the changing water trophic condition. Mixotrophic ciliates dominate in spring and summer, reaching from 58–100% of the ciliates in oligotrophic waters [23–25], but represent only 5% of the total community of ciliates in winter, probably due to the lower water temperatures and nutrients. These conditions restrict the growth of algae, negatively affecting the population of mixotrophic ciliates if their preferred species of algae is missing [24].

The mixotrophic ciliates are mainly from the genera *Mesodinium rubrum* (*Myrionecta rubra*), *Strombidium* spp., *Laboea*, *Lohmaniella,* and *Tontonia.* All of them represented by small species (30–50 μm) [23, 25, 26]. Even *Mesodinium rubrum* and other functionally photoautotrophic ciliates can sometimes contribute significantly to primary production in lakes and oceans [27]. Other species of mixotrophic ciliates are larger; for example, the genus *Stentor* is a "large" cell ~200 μm and is contributed between 49% and 68.8% of the total biomass of zooplankton in the oligotrophic lake at the Northern Patagonia of Chile [28]. *Stentor niger* represented 90% of the total ciliates biomass in Lake McCloud [29] and was the dominant protist of acidic oligotrophic lakes [30]. Some species of the genus *Stentor* also contributed with more than 50% of the plankton's photosynthesis of oligotrophic Australian lakes [31]. Dominance of this trophic group may be influenced by the limiting conditions for phototrophs, to achieve the same productivity that mixotrophs may obtain by feeding both ways. Grazing allows mixotrophs greater flexibility for balancing the supply and demand of scarce nutrients [32], a clear advantage in times of scarcity [19]. Due to their flexible nutrition, mixotrophic protists dominate in mature or more stable systems (e.g., during mild summer, in established eutrophic systems, and in oligotrophic systems). Furthermore, climate change can be expected to favor mixotrophs in the more stable water columns [32].

Ward and Follows [33] performed a global simulation of the ocean-surface food web, revealing that mixotrophy enhances the transfer of biomass to larger organisms at higher trophic levels, which in turn increases the efficiency of oceanic carbon storage through the production of larger and faster sinking conglomerates of organic molecules. It follows that mixotrophic protists play a key role in modulating the primary production that underlies the food web in aquatic systems [21, 22, 32]. However, their importance has not been fully appreciated because traditional field and laboratory studies focus on strict classifications as phototrophs or phagotrophs [32] because incorporating this flexibility to acquire food is difficult to modelize. Mixotrophy is known to be common in all aquatic systems but its contribution to net community production is difficult to quantify, and the integration of their impact on the global biogeochemical cycles remains to be incorporated.

2.2 Bacterivores

Ciliates and flagellates are the most dominant bacterivores among the phagotrophic protists in most aquatic systems [16, 34], consuming between 25–100% of the daily production of marine phytoplankton together with large quantities of bacterial biomass [18]. Bacterivores and algivorous protists are the core consumers of microbial biomass in aquatic food webs [16, 17] regulating these groups in two apparently contradictory ways: by feeding on the abundant food source, they keep in check their further expansion, that in turn gives other less preyed species the opportunity to become more numerous, and at the same time, the release of cellular wastes (from protists) enhance the reproduction of the species being predated. The combined effect of these two processes enhances the nutrient cycling and fuels biomass productivity. By performing this activity, ciliates and flagellates increases their own biomass, attracting metazoan predators and functioning as linkage of lower and upper trophic levels in aquatic food webs [16, 35, 36].

The size of the ciliate determines the sizes of preys they can feed on. Thus, bacterivorous ciliates ingest a different particle size range; the preferred size spectrum for each species is a function to cytostome size and morphology. For example, small ciliates usually bacteria eat <3 μm [18, 37, 38]. Ciliates that feed on the smallest particles (<1 μm) require relatively high densities of these bacteria as a minimum to keep their population growth [30]. Several groups of ciliates actively feed on specific bacteria species for a period ranging between 44% and 100% of the time, because bacterial densities will have variations as responses to predation intensity along time [36].

Bacterivorous ciliates are present in all aquatic environments, from oligotrophic to eutrophic, in both freshwater and oceans. The diversity of bacterivorous ciliates and their contribution to the flow of energy in trophic networks depend on the dynamics of the systems in which they are living. Therefore, food resources are probably the main regulators of ciliated communities (diversity, abundance, and biomass) [30]. For example, bacterivorous ciliates contribute very little for the direct transfer of bacterial production to the trophic networks of metazoans in oligotrophic environments. Ciliates consume less than 11% of bacterial productivity in these waters [39–41]. Perhaps the heterotrophic bacteria that are very small in these lagoons (0.035 to 0.4 μm) are grazed by bacterivorous ciliates at a very low rate [41], or the number of bacteria is not enough to support larger ciliate communities feeding on smaller bacteria (<1 μm), as they require high densities of bacteria to maintain their populations [30]. Then, productivity of oligotrophic systems function most of the time as bottom-up (availability of substrate and nutrients) controlled [42]. This functioning will remain until seasonal pulses of nutrients (or human subsidies) arrive, busting primary productivity and changing the system into top-down control, and it will keep functioning the same way until the pulse of nutrients (or subsidy) is completely metabolized, returning the system to the bottom-up dynamic.

Contrastingly, densities of heterotrophic bacteria in eutrophic environments are sufficiently higher to also keep a higher diversity of active bacterivores [43], fueling ciliates biomass productivity and allowing the intervention of metazoan predation. Top-down control (predation) seems to be in function all the time for regulating the abundance of heterotrophic bacteria in eutrophic systems [42]. Normally, communities of bacterivorous ciliates of small sizes (~ 30 μm) are found as dominant in eutrophic environments [30, 38]. The most abundant ciliates in these environments are small oligotrichs (*Halteria*), scuticociliates (*Cyclidium*), and Peritrichs (*Vorticella*) [30, 38, 44, 45]. *Halteria grandinella,* for example, is one of the most important bacterivores due to its high consumption rate of bacteria [38], the genus *Halteria* is very important in meso-eutrophic lakes because they prey on a large spectrum of sizes, from bacterial cell measuring just around 0.4 μm to up to 5 μm; they have a high potential growth rate, because of its efficient nutrient absorption, and show defensive strategies reducing their vulnerability to predation by metazooplankton in comparison to other common pelagic ciliates [45].

Sessile ciliates such as *Vorticella* and *Epistylis* are typical members of protists' community in aquatic environments [34, 45–49]. They heavily graze on bacteria having even higher ingestion rates than free-swimming bacterivorous protist and can account for 66% of total bacterivores. Even in small numbers, epibiotic ciliates can have a great grazing impact on bacteria [34]. For example, *Vorticellides aquadulcis* had the highest grazing rates of all the ciliates from a meso-oligotrophic lake community [38]. Some common bacterivorous ciliates are found in **Table 1**.

2.3 Feeding on phototrophs

There is a difficulty in assessing a proper name for the kind of food protists feed on when they become predators of phototrophs, as this group consists of both eukaryotic and procaryotic members, and neither of these primary producers may be considered as "plants" or "herbs". Feeding on them cannot be considered as herbivory. On the procaryotic part, cyanobacteria are a phylum comprising many species that, besides being phototrophs, can also produce toxic molecules, compromising the fresh water supplies for human consumption when growing unchecked in oligotrophic waters [50, 51]. From the eukaryotic part, there is an extra complication when trying to separate the permanent phototrophs from the mixotrophs.

Moving the sizes up, ciliates are one of the most important groups feeding on phytoplankton in marine and freshwater environments [18, 41, 52]. They may consume up to 74% of the daily phytoplankton production [53], becoming the key controllers of phytoplankton biomass [54]. On the other hand, ciliates mobilize the

Trophic groups	Examples	References
Bacterivores	Colpodida (*Colpoda*), Peritrichia (*Carchesium*, *Epistylis*, *Vorticella*), Scuticociliatia (*Cyclidium*, *Parauronema*, *Pseudocohnilembus*), Stichotrichia (*Halteria*)	[34, 46, 48, 114, 116]
Feeding on Phototrophs	Choreotrichia (*Codonella*, *Strobilidium*), Oligotrichia (*Pelagostrombidium*) Heterotrichea (*Linostomella*), Peniculia (*Frontonia*), Tintinnida (*Helicostomella*, *Ptychocylis*, *Tintinnidium*, *Tintinnopsis*)	[54, 56–58, 60]
Predators of predators	Heterotrichea (*Fabrea salina*) Litostomatea (*Didinium*, *Lacrymaria*, *Lagynophrya*, *Loxophyllum*, *Mesodinium*, *Monodinium*, *Phialina*) Prostomatea (*Balanion*, *Holophrya*, *Tiarina*), Stichotrichida (*Halteria*)	[62, 64, 66, 68, 72, 108]
Omnivorous	Choreotrichia (*Rimostrombidium*), Hypotrichia (*Euplotes*), Prostomatea (*Urotricha*, *Coleps*), Scuticociliatia (*Pleuronema*), Stichotrichida (*Oxytricha*, *Stylonychia*)	[49, 66, 69, 73, 79]
Mixotrophos	Litostomatea (*Mesodinium rubrum*), Oligotrichia (*Laboea*, *Strombidium*, *Tontonia*), Choreotrichia (*Lohmanniella*), Heterotrichea (*Stentor*)	[23, 25, 26, 28, 29, 31]

Table 1.
Trophic groups free-living ciliates in aquatic environments.

highest proportion of organic carbon and nutrients in oligotrophic waters dominated by cyanobacteria, playing the fundamental role of linking the productivity of microbial food web with the metazoans [41, 53]. It has also been noticed that the flux of carbon up to metazoans is not interrupted when the density of bacterivores ciliates falls, but it is compensated by predation on ciliates feeding on phototrophs [41]. Some of the ciliates that feed on phototrophs are in **Table 1**.

Ciliates feeding on phototrophs represent between 30–65% of the total biomass of all functional groups of ciliates thriving in eutrophic lakes [55]. However, this dominance is not permanent. Ciliates feeding on phototrophs become very numerous on the blooming season [56], and even dominate the entire ciliate community for short periods between seasons [57].

Tintinnids tend to feed on small-cell-sized phytoplankton (2–20 μm) [58]. They are voracious phytoplankton feeders that may consume over half the quantity of these kind of phototrophs in marine waters [54] and over 69% of these primary producers in lakes [59]. Species like *Helicostomella subulata*, *Ptychocylis* spp., and *Parafavella* spp. make a significative contribution to biomass of ciliates feeding on phototrophs in marine environments [60].

Selective feeding has been observed in several species of ciliates. However, feeding on a wider spectrum of sizes and kind of phototrophs (non-selective feeding) allows them to take advantage of the productivity in hypereutrophic environments rich in small particulates [49]. The genus *Tintinnidium* groups species that dwell very well in these kinds of waters and may be used as model organisms to study the ciliates' adaptation to excess of organic matter [61].

2.4 Predators of predators or raptorial feeders

There are several species of ciliates and flagellates that feed on bacterivorous protists and on protists feeding on phototrophs. These are predators of predators. These predator species may feed temporarily on bacteria but cannot survive by just this consumption; they are attracted to them as they offer clues to discover their preferred preys: ciliates, flagellates, or amoebae feeding on bacteria.

Most of predator ciliates feed on preys around 10 times smaller than them [62, 63], although raptorial feeders may consume bigger preys, comparable to their own size or even bigger [64]. This capacity is due to their very flexible cytostome as is the case in protostomatids genera *Tiarina*, *Balanion,* and *Holophrya,* and in the litostomatid genus *Didinium* [62]. Ciliates select their food based on prey's size, motility, and biochemical composition of cells' surface [62]. Raptorial ciliates exert strong pressure on populations of small phototrophic and heterotrophic flagellates [65], imprinting some velocity to nutrient cycling in environments where productivity allows them to develop large populations.

Predatory ciliates are present in small numbers along seasons in oligotrophic waters, showing surges in population numbers, in synchrony with the increase of primary productivity during the spring, reaching up to 55% of the total ciliates' abundances in temperate waters [64, 66]. However, they only reach between 24.6% to 28.7% in freezing oligotrophic waters of the Arctic and Antarctic [67].

On the other hand, predatory ciliates become important top-down controllers of microbial food web productivity in eutrophic and hypertrophic waters [68]. Eutrophic waters have the conditions to sustain high productivity rates of phototrophs and heterotrophic bacteria, sustaining, in turn, large populations of their grazers, promoting the increase of predatory ciliate population [69]. Biomass of raptor ciliates may reach almost an order of magnitude higher in eutrophic compared to the one obtained in meso and oligotrophic lakes, suggesting that they are effectively controlling the primary productivity [70]. This assumption is supported by the covariance of predatory ciliates and their preferred food. For example, the increasing population of predatory ciliates bigger than 100 μm is related to a simultaneous shrinkage of abundance of smaller ciliates (<20–40 μm), mostly phototrophs and bacterivorous [71]. Big and voracious ciliate raptors like *Monodinium* sp. and *Lagynophrya* sp. have stronger impact than rotifers on populations of small ciliates [68]. However, quantity of prey is not the only important factor, and species diversity is needed to sustain more raptorial species of ciliates. For example, only *Monodinium* remained abundant when diversity of preys falls below a limit [72].

Several species of oligotrichs feed on bacterivorous flagellates, showing an efficiency of 45% biomass transformation, also fueling the bacterial productivity by releasing essential nutrients for heterotrophic bacteria to keep their population growth [65]. Some predatory ciliates are shown in **Table 1**.

2.5 Omnivorous

Omnivorous protists are an important group to look for when assessing the stability of a food web because their very presence means productivity is enough to non-specialists, to feed on a variety of resources. Omnivores strengthen the resilience of planktonic communities by regulating the trophic dynamics [73]. Omnivorous ciliates may have a preferred prey but can easily move to other kinds of prey, which may be more abundant or easier to catch [74]. This variety of resources for true omnivorous ranges from bacteria, algae, other ciliates of different sizes to fungi [73]. This versatility gives them an advantage to withstand a resource limitation by having alternative prey [70]. Additionally, omnivorous ciliates increase the stability of planktonic communities by feeding on species that may pass undetected from their specialized predators, by having densities small enough to get an advantage of the elimination of their competitors and increase their numbers. In this situation, omnivores would prevent them to reach high densities too fast, giving time for their specialized predators to increase to population levels that may effectively control the newly abundant prey.

Omnivorous ciliates are present in any kind of environment allowing the stability of protist communities. They are elements of marine and freshwater ecosystems, both oligotrophic [66, 75] and eutrophic [69, 76], as well as in polar waters [67].

As with the other trophic groups, omnivores show seasonal bursts of abundances in the communities they are part of, especially in oligotrophic waters where they are scarce most of the time, except for occasional bursts [77, 78]. Omnivorous ciliates are commonly found in lakes throughout the year, normally with low species richness, representing between 2–12% of the ciliates species [67, 79]. Their low contribution to the number of individuals makes them reach a peak of 35% during productivity bursts [66, 79]. However, this proportion may steadily increase in the proportion the environment is turning into the eutrophic condition, increasing the species richness, although the densities of omnivorous ciliates may momentarily diminish with the eutrophication [69] as result of the species increase (more species and lower number of individuals by species). Once the eutrophication reaches a steady state, the biomass of the omnivorous ciliates will reach high values and even dominate among ciliates [76].

The numbers of small omnivorous ciliates usually dominate in meso oligotrophic environments, feeding on dominant bacteria (<2 μm) and algae (2–20 μm) [49]. Food concentration is a very important factor, strongly affecting an easily detectable feeding behavior of omnivorous ciliates [73]. Several of the most common omnivorous ciliates are shown in **Table 1**.

3. Boundaries among trophic webs. Is that possible?

Functionality alone has its own complexity in food webs because, for example, mixotrophs would be functioning as phototrophs or as heterotrophs along different hours during the same day (How long do they function as phototrophs? How long do they the function as heterotrophs?). An extra dimension in this world comes from the different sizes of preys corresponding to the predators' sizes and the number of cells each individual predator must get to produce another individual [80]. This is one of the reasons why plankton has been divided in microplankton, nanoplankton, and picoplankton. Each category corresponds to the range sizes of microorganisms. The smaller ones like picoplankton and nanoplankton, performing primary productivity (chemolithotrophs or phototrophic [3], can sustain their corresponding predator's size and be up to ten times bigger, namely nanoflagellates and microflagellates. These are the two groups of protists related to their size and morphology rather than their taxonomic affiliation [81], since very few information is known about them apart from 18S SSU rDNA sequences; they have been recognized performing predatory activity on phototrophs of the smaller sizes.

One alternative to conceptually reduce the complexity of microbial food webs is analyzing them as nested compartments. This means that the transfers of matter and energy takes place inside each compartment corresponding with one size class of producers and its predators because these organisms function in the same time frame. Then, several of these compartments may get integrated in a bigger one by predation of the next size class. Time frame for this bigger class is also bigger than the previous one, as the sizes of the organisms are also bigger and so on. Every compartment of bigger sizes function as concentrator of biomass and disperser of energy. However, the wastes generated in each compartment releases the nutrients once fixed in the biomass fueling the nutrient cycle in compartments of all sizes. Up to here, it looks like the aquatic food web is functioning as a continuum along and through the water column and surface. However, there is a chance of recognizing boundaries to help a better understanding the food webs dynamic.

When hearing the word "boundary", immediately, the existence of physical barriers delimiting something in space comes to mind. Because of that, it is hard to imagine an aquatic food web being physically limited because our experience has shown us the big animals feeding on all planktonic organisms at once, which could be in thousands or even millions. However, it just represents a small appetizer for a whale.

A careful examination reveals that very small organisms live faster than ones at the immediate upper-sized scale and intuition tells us that time may be experienced in different ways, depending on the size of organisms involved. The size ranges occupied by ciliates in the microbial food web spans from less than 10 μm to more than 4500 μm [82]. Comparatively, their pool of size ranges would be like the pool of sizes from small fishes to whales. Why are these sizes important? Because it can be argued that the velocity of nutrient exchange is faster in the smaller organisms and the nutrients may be "sequestered" for long periods by the bigger and long-lasting animals. In this way, a complication of time arises when trying to diagram the nutrient cycle in the microbial food web. Time becomes another varying feature rather than a constant in food web dynamics. In this way, time may draw the boundaries between compartments and, at the same time, could be avoiding contradicting the nested compartment proposal in the physical limitless aquatic system.

4. The soil system

It is easier to recognize physical boundaries in terrestrial ecosystems as the environment changes at slower velocities than the very dynamic aquatic environment. Soil is a heterogenic environment, the opposite to the aquatic ones. It is an environment that cannot be seen through and be dived in. Soil matrix is composed of a very complex mixture of mineral particles, organic matter and living organisms. This mixture is organized in aggregates that may facilitate or resist water and air passing through it but, most importantly, these aggregates proportionate spaces where all living beings can move through soil.

At a microscale, soil aggregates divide the open spaces in two types, the fast water passing by (the space between aggregates) and the slow motion of water in the space inside the aggregate, and consequently of slow-moving air too, as air and water move through the same spaces). These are the soil's physical boundaries, and this is the environment where roots move and look for hotspots of nutrients, as well as places where microbial symbionts may be found (normally inside soil aggregates). Water reaching soil aggregates dissolved salts and polar molecules that may contain nutrients that will be taken by roots, mycorrhiza, or bacteria. This is a complementary start of plants primary productivity, because plants have to take water from soil together with other nutrients to produce a wide range of molecules, from non-protein forming amino acids to scents and pheromones, as result of what is known as the "secondary metabolism." Plant primary productivity comprises both photosynthesis-respiration (primary metabolism) and secondary metabolism, irrespective of being vascular or nonvascular.

Soil productivity is dependent on the nutrient exchange velocity rather than the gross amount of bioavailable nutrients. Nutrients used and released very fast means energy is being captured, transformed, and degraded very fast, implying the activities of all participating organisms are taking place so fast that production of biomass at all levels is gaining momentum and its control may come only from consumption (top-down) no matter that nutrients exist in limited quantity. This feature also explains why the smaller organisms can sustain productivity of the biggest ones. In other words, aerial part of plants are very important for primary productivity because it is the place were light, inorganic carbon, and water are used to produce organic molecules that are at the base of primary productivity (Sun's energy fixation in organic molecules).

Without diminishing photosynthesis' importance, most of terrestrial plants gather a "productivity teamwork" inside and around their roots, involving mycorrhizal fungi and mutualistic bacteria, a functional place known as the rhizosphere. Almost 80% of the known terrestrial plants need the association with a mycorrhiza, to appropriately complete their life cycle, but all plants need mutualistic bacteria to grow. Microbial partners are indeed an important part of primary productivity, as they actively participate in the acquisition, modification, and metabolization of many organic molecules containing the elements we call "Nutrients." For example, it has largely been demonstrated that mycorrhiza translocate phosphorus to plants. At present, very few people challenge this. However, what form of phosphorus is translocated from mycorrhiza to plant? Surely, it is not the phosphorous as molecule, but organic molecule where P is forming part of the structure. Plants can take up P from inorganic molecules in general or from phosphoric acid. Why do they need mycorrhiza to supply P? It is still an open question, but the degree of specificity of the plant-mycorrhiza association allows to conjecture that plant and mycorrhiza share metabolites containing nutrients (not just P) for metabolic complementation, and the same could be true for mutualistic bacteria. This would explain why one species of mycorrhizal fungi is mutualistic to several plant species but functions as pathogenic or parasite to other ones.

Contrary to what happens in waters, soil fungi and bacteria are scattered through soil and physically constrained to available surfaces. If they keep growing unchecked, bacteria may become effective nutrient competitors to plants, as nutrients forming bacterial biomass are non-available to plants. Mycorrhiza may move farther away from the root than bacteria and can establish a mutualistic relationship with other roots (whether they are from the same plant or from a different species, it does not matter) to avoid becoming competitors. Absence of bacterivores is a needed condition for bacteria to become a plant competitor in the rhizosphere [83, 84]. Bacterivores ciliates, flagellates, and amoebae release nutrients trapped in bacterial biomass, stimulating both plant and bacterial growth. In the first case, nutrient release allows roots to take them in and bacteria microcolonies may grow again in the root surfaces, already cleaned out, and obtain nutrients from predators' wastes [84].

Soil's physical constrains allow growth of bacteria and fungi in differentiated places. Sometimes bacteria also grow on the surface of hypha, helping fungi to mimic bacteria and somehow escape from fungal predators. It has been possible to observe protists feeding predominantly on fungi and avoiding bacteria as much as possible (*Dermamoeba granifera*, *Cochliopodium* sp.). There are also protist species feeding on soil algae (*Colpoda* sp., *Polychaos* sp., *Thecamoeba* sp.) Consequently, it is possible to recognize the existence of several functional groups of soil protists: few species of phototrophs feeders, large quantity of bacterivores, fungal feeders, raptorial feeders (*Balamuthia mandrillaris*), and omnivores (*Acanthamoeba castellanii*, *A. polyphaga*, *A astronyxis*).

This differentiation of soil's physical spaces makes it easier to visualize the small productivity compartments around roots, absorbing hairs inside small soil aggregates, bigger compartments covering aggregates on the tip of the root and getting in contact through fungal hypha.

Motility of bigger protists are limited to litter and upper soil layer by the available spaces, restricting their abundance in the underneath layers. Testate amoeba, ciliates, and flagellates, around 100 μm, dominate in these 2 layers and actively participate nutrient recycling from litter, while smaller size ciliates like *Colpoda cucullus*, small flagellates and small naked amoebae distribute better in the underlying soil strata in and around soil aggregates.

Primary productivity in soil is restricted to the upper layers where cyanobacteria and eukaryote algae may survive and even form thin layers known as microbial soil crusts. Both phototrophic bacteria and algae may form stable mutualistic symbiosis with other organisms, like fungi, to develop thicker structures composing soil crusts showing lichens and mosses. Beneath and into soil crusts, ciliates, flagellates, and amoebae are among the most important microbial predators, active mainly during the time of water availability [85, 86]. However, the main photosynthetic carbon input is released by roots into soil layers [87]. Roots secrete amino acids and other complex organic molecules to attract symbiotic bacteria and mycorrhiza conforming the trio of soil productivity sustaining microbial food webs deep into soil [88, 89]. Consequently, protists' species diversity may be higher around roots and the dominance of ciliates may be restricted to the sizes of soil pores [86, 90–92]. Soil protists were recognized as purely bacterivorous because fungi feeding protists may transitorily feed also on bacteria. However more detailed studies have recognized species of soil protists feeding only on bacteria or fungi [93–95]. Among the main bacterivorous ciliates are Colpodida (*Breslaua vorax*, *Colpoda aspera*, *Colpoda inflata*, *Colpoda maupasi*, *Colpoda steinii*, *Cyrtolophosis elongata*, *Cyrtolophosis mucicola*, *Platyophrya vorax*, *Pseudocyrtolophosis terricola*, *Pseudoplatyophrya nana* [85, 96].

Fungi and bacteria normally use different kind of organic molecules, bacteria normally metabolize low molecular weight organic molecules while fungi normally metabolize complex organic polymers of high molecular wight [97]. This metabolic difference allows to conceptualize two pathways for nutrient cycling: the bacterial and the fungal paths. However, this concept is being challenged because of the abundance of protists feeding on both kind of microorganisms [98, 99]. All the early recognized fungi feeding ciliates and amoebae in soil ranges from 50 microns to above 150 μm [100]. However, there are also smaller ciliates and flagellates feeding on both spores and hypha [100]. The main groups of specialized fungal feeder ciliates are grouped in the family Grossglockneriidae [93]. This family of ciliates may account for mora than 2% of the protists sequences in the forest litter and grassland while may drop below 0.3% in peatland soil, probably due to the reduction of soil pore sizes [100]. Although, counting techniques based in MPN calculated around 200 cells/gram soil DW in previous studies [101]. Protists have a very limited capacity to disperse throughout the soil system by themselves. However, oligochaeta disperse them as cysts farther than a few centimeters, in the range of several meters both horizontally as well as vertically into the soil system.

Soil functioning is much more variable than the aquatic systems, as it is regularly subjected to dryness and several flooding events per year. For microbial ecologists, soil is a natural stressed environment, having enormous variations of water availability through seasons, especially in arid and semiarid environments. However, there is a comparable situation, although at lesser degree, in the tropical dry forests, temperate, and tundra regions. Even at the equator, the rainy forests may show an excess of soil in water, stressing microbial food webs.

DOI: http://dx.doi.org/10.5772/intechopen.97252

5. Perturbations and food webs

Microbial communities have been evolved by modifications and adaptations in responses to natural stresses that finally allow them to get along with environmental change. The problem we are facing now resides in the velocity of environmental changes imprinted by human activities. The most important, but hardly the only one, resides in the use of fossil fuels because of the acceleration of climate change. The CO2 released as byproduct of combustion is just one of the causes of climate modification in the short term (in historical and geological times). Internal combustion engines also produce other greenhouse gases such as NxO or NO2, having a bigger capacity of keep heat, and this is a big problem generated only for the atmosphere. Hydrocarbons pose a permanent threat of contamination to aquatic and soil systems near the extraction zones, the transporting infrastructure to refineries, infrastructure for later transportation as fuel to expending places, and by illegal activities damaging oil ducts.

Soil microbiota react in different ways along the gradient of contamination when hydrocarbons reach soils. The plume of contamination normally eradicates the phototrophs and exert a strong selective pressure on bacteria and fungi, by killing or inhibiting the growth of sensitive species while enhancing the growth of resistant ones. These effects can be modified by the toxicity of the different compounds rupturing and/or changing the connections of the trophic networks [102, 103].

The effect of hydrocarbon contamination and others contaminants (pesticides, heavy metals) on communities will depend on the intensity, duration, and frequency of the perturbation. Then, lower species richness and abundance, shortening of the trophic webs, and the simplification of the trophic web are among the first observable damages contamination cause on microbial and protist communities [104]. Protists must at least tolerate the presence of the contaminant to achieve this function. Protists do not feed on hydrocarbons, but their grazing activity on the microorganisms that can keep the metabolization of the contaminant as high as another limiting factor allows them to.

Greater richness and abundance of ciliates species are associated with less perturbed areas; the greater the perturbation, the lesser species richness and abundance [105], regardless of the nature of the perturbing factor. For example, a significant reduction of ciliate diversity has been found in systems polluted by high hydrocarbon concentrations [106]. Medium concentrations only reduce the quantity of individuals from dominant species [106], while low concentrations produce an increase in the numbers of heterotrophic protists [107]. Saline accumulation forces the ciliates' diversity to decrease as salinity values increase [108, 109]. In the same way, acidic pollution produces lower species richness and abundance as the environment becomes more acidic [110, 111], and the same pattern is observed with heavy metals' contamination [104, 110].

Addition of organic matter in excess suddenly changes the base of production of the microbial food web, from phototrophs' productivity to heterotrophic bacteria and yeasts' productivity. The time of reaction is also different along the different microbial groups surviving the contamination event. Bacteria may start their biological activities several hours after the pollution event, whereas yeast and protists will delay from days to weeks, depending on the size of the organism.

Changes of primary producers from phototrophs to heterotrophs scale to functional groups, accommodating species richness and abundance of bacterivores protists, followed by omnivores. This is due to hydrocarbons stimulation of bacterial growth and the consequently increase of bacterivores species [112, 113]. Some species of genera *Colpoda* and *Vorticella* dominate aquifers receiving constant hydrocarbon discharges [114]. The bacterivorous ciliates, *Parauronema virginianum,* strongly dominate sites highly polluted with hydrocarbons and are replaced by *Pseudocohnilembus* and *Euplotes* later [115]. Additionally, organic contamination and heavy metals increase the abundance of bacterivorous ciliates in water and sediments [116].

An increase in diversity and complexity of food webs are direct effects of these perturbations. Oil spill in deep waters increase the richness of the microbial community species and the complexity of their corresponding relationships, and the oil stimulated microbial activity supports greater variety of ciliates functioning along several trophic levels [117].

Other events of enriching oligotrophic systems with organic matter produce similar changes in the community structure of ciliates. Tirjaková and Vďačný [118] analyzed the changes in the communities of ciliates before and after a windstorm hit a stream, and they found a significant increase of ciliates' species' richness and abundance after the storm. Several weeks later, the community of ciliates presented the typical values of oligotrophic sites. The increase in resources availability is the factor indirectly responsible of these changes of ciliate community, but later, communities tend to return to states similar to the initial ones after resources exhaustion, which my take place around six months [118]. However, Shabarova et al. [119] report that the microbial community recovers from perturbation to a pre-flood state within two weeks after the event.

Regarding the connections' shrinkage of the trophic networks, a gradual narrowing of the planktonic size spectrum has been reported in hypersaline lakes, correlated to salinity increases during the summer, resulting in a simplification of the community represented by the ciliated *Fabrea salina*, diatoms, and *Dunaliella* spp. [120]. Simplifications of food webs have also been described as consequence of heavy metal contamination, herbicide use, and lake acidification [104, 121, 122]. Loss of connections have consequences on carbon transfer in food webs. The decrease of bacterivores species allows an excessive increase in bacterial biomass, which may produce up to 300-fold reduction in the transfer of carbon from the bacteria to higher levels of the trophic networks [104].

Communities' characteristic of hypersaline lakes are dominated by *Fabrea salina,* which has a broad tolerance to salinity and contributes to high proportion of the biomass of ciliates in hypersaline lakes [108, 109, 120, 123]. In addition, its abundance is strongly related to the microalgae, *Dunaliella* sp. [123], and can act as a competitor to shrimp, *Artemia salina*, in saline environments [108].

Regarding the perturbances in the soil ciliated communities, similar effects have been described as in aquatic ecosystems. Exposure of ciliate communities to heavy metals induces a reduction in the biomass of ciliates and this effect lasts for 20 weeks [124]. Insecticides also generate a decrease in ciliates species immediately after contamination, they also generate a change in the dominance of ciliates, the bacterivores (*Colpoda* spp. and *Paracolpoda steinii*) and macrophage (*Grossglockneria*) considerably increased their abundance after 90 days, while that other genera of ciliates decreased [125]. In soils contaminated with hydrocarbons, a decrease in diversity and a lower functional diversity have also been observed, the ciliated communities in soils with hydrocarbons are dominated by the Colpodea class [96, 126, 127]. It has also been observed that along with the decrease in the diversity of ciliates there is a decrease in the trophic groups after an intense pulse of contamination by hydrocarbons. However, the community recovers its diversity and trophic groups after a month of contamination [127].

6. Conclusions

Protists in general, and ciliates in particular, play a key role in nutrient cycling and food web functioning in both aquatic and terrestrial ecosystems. In the world experiencing climate change and other kind of anthropogenic menaces, protists may be useful partners to tell us how aquatic and terrestrial systems are dealing with these issues while mesmerizing the observer with their great diversity of beautiful forms.

Conflict of interest

The authors declare no conflict of interest.

Author details

Laura Mondragón-Camarillo* and Salvador Rodríguez Zaragoza
Laboratorio de Ecología Microbiana, UBIPRO, FES-Iztacala UNAM, México, México

*Address all correspondence to: zombo_7@yahoo.com.mx

References

[1] Johnson CA, Bronstein JL. Coexistence and competitive exclusion in mutualism. Ecology. 2019;100(6):e02708. doi:10.1002/ecy.2708

[2] Kneitel JM. Gause's Competitive Exclusion Principle. In: Encyclopedia of Ecology. Elsevier; 2019:110-113. doi:10.1016/B978-0-12-409548-9.00816-2

[3] McCollom TM. Geochemical constraints on primary productivity in submarine hydrothermal vent plumes. Deep Res Part I Oceanogr Res Pap. 2000;47(1):85-101. doi:10.1016/S0967-0637(99)00048-5

[4] Felbeck H, Somero GN. Primary production in deep-sea hydrothermal vent organisms: roles of sulfide-oxidizing bacteria. Trends Biochem Sci. 1982;7(6):201-204. doi:10.1016/0968-0004(82)90088-3

[5] Hui D. Food Web: Concept and Applications. Nat Educ Knowl. 2012;3(12):6.

[6] Elton CS. The Ecology of Invasions by Animals and Plants. Springer US; 1958. doi:10.1007/978-1-4899-7214-9

[7] Fuhrman J. Bacterioplankton Roles in Cycling of Organic Matter: The Microbial Food Web. In: Primary Productivity and Biogeochemical Cycles in the Sea. Springer US; 1992:361-383. doi:10.1007/978-1-4899-0762-2_20

[8] Johnson S, Domínguez-García V, Donetti L, Muñoz MA. Trophic coherence determines food-web stability. Proc Natl Acad Sci U S A. 2014;111(50):17923-17928. doi:10.1073/pnas.1409077111

[9] Singh BN. Toxic effects of certain bacterial metabolic products on soil Protozoa [2]. Nature. 1942;149(3771):168. doi:10.1038/149168a0

[10] Singh BN. The selection of bacterial food by soil amoebae, and the toxic effects of bacterial pigments and other products on soil protozoa. Vol 26. Wiley-Blackwell; 1945.

[11] Doyle WL. The nutrition of the protozoa. Biol Rev. 1943;18(3):119-136. doi:10.1111/j.1469-185X.1943.tb00293.x

[12] Provasoli L. Nutrition and Ecology of Protozoa and Algae. Annu Rev Microbiol. 1958;12(1):279-308. doi:10.1146/annurev.mi.12.100158.001431

[13] Zengler K, Zaramela LS. The social network of microorganisms - How auxotrophies shape complex communities. Nat Rev Microbiol. 2018;16(6):383-390. doi:10.1038/s41579-018-0004-5

[14] Nichols D, Lewis K, Orjala J, et al. Short peptide induces an "uncultivable" microorganism to grow in vitro. Appl Environ Microbiol. 2008;74(15):4889-4897. doi:10.1128/AEM.00393-08

[15] Strom SL, Loukos H. Selective Feeding by Protozoa: Model and Experimental Behaviors and Their Consequences for Population Stability. Vol 20.; 1998.

[16] Azam F, Fenchel T, Field J, Gray J, Meyer-Reil L, Thingstad F. The Ecological Role of Water-Column Microbes in the Sea. Mar Ecol Prog Ser. 1983;10:257-263. doi:10.3354/meps010257

[17] Caron DA. Protistan herbivory and bacterivory. Methods Microbiol. 2001;30:289-315. doi:10.1016/s0580-9517(01)30050-8

[18] Sherr EB, Sherr BF. Bacterivory and herbivory: Key roles of phagotrophic protists in pelagic food webs. Microb Ecol. 1994;28(2):223-235. doi:10.1007/BF00166812

[19] Weisse T, Sonntag B. Ciliates in planktonic food webs: Communication and adaptive response. In: Biocommunication of Ciliates. Springer International Publishing; 2016:351-372. doi:10.1007/978-3-319-32211-7_19

[20] Mitra A, Flynn KJ, Tillmann U, et al. Defining Planktonic Protist Functional Groups on Mechanisms for Energy and Nutrient Acquisition: Incorporation of Diverse Mixotrophic Strategies. Protist. 2016;167(2):106-120. doi:10.1016/j.protis.2016.01.003

[21] Moeller H V., Peltomaa E, Johnson MD, Neubert MG. Acquired phototrophy stabilises coexistence and shapes intrinsic dynamics of an intraguild predator and its prey. Ecol Lett. 2016;19(4):393-402. doi:10.1111/ele.12572

[22] Stoecker D, Johnson M, deVargas C, Not F. Acquired phototrophy in aquatic protists. Aquat Microb Ecol. 2009;57(3):279-310. doi:10.3354/ame01340

[23] Bernard C, Rassouladegan F. Seasonal variations of mixotrophic ciliates in the northwest Mediterranean Sea. Mar Ecol Prog Ser. 1994;108(3):295-302. doi:10.3354/meps108295

[24] Holen DA. The relative abundance of mixotrophic and heterotrophic ciliates in an oligotrophic lake. Arch fur Hydrobiol. 2000;150(1):1-15. doi:10.1127/archiv-hydrobiol/150/2000/1

[25] Mironova E, Telesh I, Skarlato S. Planktonic ciliates of the Neva Estuary (Baltic Sea): Community structure and spatial distribution. Acta Protozool. 2013;52(1):13-23. doi:10.4467/16890027AP.13.002.0830

[26] Onda DFL, Medrinal E, Comeau AM, Thaler M, Babin M, Lovejoy C. Seasonal and interannual changes in ciliate and dinoflagellate species assemblages in the Arctic Ocean (Amundsen Gulf, Beaufort Sea, Canada). Front Mar Sci. 2017;4(FEB):16. doi:10.3389/FMARS.2017.00016

[27] Esteban GF, Fenchel T, Finlay BJ. Mixotrophy in Ciliates. Protist. 2010;161(5):621-641. doi:10.1016/j.protis.2010.08.002

[28] Woelfl S. The distribution of large mixotrophic ciliates (Stentor) in deep North Patagonian lakes (Chile): First results. Limnologica. 2007;37(1):28-36. doi:10.1016/j.limno.2006.08.004

[29] Bienert RW, Beaver JR, Crisman TL. The Contribution of Ciliated Protozoa to Zooplankton Biomass in an Acidic, Subtropical Lake. J Protozool. 1991;38(4):352-354. doi:10.1111/j.1550-7408.1991.tb01372.x

[30] Beaver JR, Crisman TL. Analysis of the community structure of planktonic ciliated protozoa relative to trophic state in Florida lakes. Hydrobiologia. 1989;174(3):177-184. doi:10.1007/BF00008155

[31] Laybourn-Parry J, Perriss SJ, Seaton GGR, Rohozinski J. A mixotrophic ciliate as a major contributor to plankton photosynthesis in Australian lakes. Limnol Oceanogr. 1997;42(6):1463-1467. doi:10.4319/lo.1997.42.6.1463

[32] Mitra A, Flynn KJ, Burkholder JM, et al. The role of mixotrophic protists in the biological carbon pump. Biogeosciences. 2014;11(4):995-1005. doi:10.5194/bg-11-995-2014

[33] Ward BA, Follows MJ. Marine mixotrophy increases trophic transfer efficiency, mean organism size, and vertical carbon flux. Proc Natl Acad Sci U S A. 2016;113(11):2958-2963. doi:10.1073/pnas.1517118113

[34] Carrias JF, Amblard C, Bourdier G. Protistan bacterivory in an oligomesotrophic lake: Importance of attached ciliates and flagellates. Microb Ecol. 1996;31(3):249-268. doi:10.1007/BF00171570

[35] Sherr E, Sherr B. Role of microbes in pelagic food webs: A revised concept. Limnol Oceanogr. 1988;33(5):1225-1227. doi:10.4319/lo.1988.33.5.1225

[36] Gonzalez. Bacterivory rate estimates and fraction of active bacterivores in natural protist assemblages from aquatic systems. Appl Environ Microbiol. 1999;65(4):1463-1469. doi:10.1128/AEM.65.4.1463-1469.1999

[37] Beaver JR, Crisman TL. The role of ciliated protozoa in pelagic freshwater ecosystems. Microb Ecol. 1989;17(2):111-136. doi:10.1007/BF02011847

[38] Šimek K, Bobková J, Macek M, Nedoma J, Psenner R. Ciliategrazing on picoplankton in a eutrophic reservoir during the summer phytoplankton maximum: A study at the species and community level. Limnol Oceanogr. 1995;40(6):1077-1090. doi:10.4319/lo.1995.40.6.1077

[39] Epstein SS, Burkovsky I V., Shiaris MP. Ciliate grazing on bacteria, flagellates, and microalgae in a temperate zone sandy tidal flat: ingestion rates and food niche partitioning. J Exp Mar Bio Ecol. 1992;165(1):103-123. doi:10.1016/0022-0981(92)90292-I

[40] Karayanni H, Christaki U, Van Wambeke F, Thyssen M, Denis M. Heterotrophic nanoflagellate and ciliate bacterivorous activity and growth in the northeast Atlantic Ocean: a seasonal mesoscale study. Aquat Microb Ecol. 2008;51(2):169-181. doi:10.3354/ame01181

[41] Sakka A, Legendre L, Gosselin M, Delesalle B. Structure of the oligotrophic planktonic food web under low grazing of heterotrophic bacteria: Takapoto Atoll, French Polynesia. Mar Ecol Prog Ser. 2000;197:1-17. doi:10.3354/meps197001

[42] Kalinowska K, Guśpiel A, Kiersztyn B, Chróst RJ. Factors controlling bacteria and protists in selected Mazurian eutrophic lakes (North-Eastern Poland) during spring. Aquat Biosyst. 2013;9(1):9. doi:10.1186/2046-9063-9-9

[43] Sanders RW, Caron DA, Berninger UG. Relationships between bacteria and heterotrophic nanoplankton in marine and fresh waters: an inter-ecosystem comparison. Mar Ecol Prog Ser. 1992;86(1):1-14. doi:10.3354/meps086001

[44] Posch T, Eugster B, Pomati F, Pernthaler J, Pitsch G, Eckert EM. Network of interactions between ciliates and phytoplankton during spring. Front Microbiol. 2015;6(NOV):1289. doi:10.3389/fmicb.2015.01289

[45] Šimek K, Jürgens K, Nedoma J, Comerma M, Armengol J. Ecological role and bacterial grazing of Halteria spp.: Small freshwater oligotrichs as dominant pelagic ciliate bacterivores. Aquat Microb Ecol. 2000;22(1):43-56. doi:10.3354/ame022043

[46] Hunt JF, Amblard C, Bourdier G. Seasonal dynamics and vertical distribution of planktonic ciliates and their relationship to microbial food resources in the oligomesotrophic Lake Pavin. Arch fur Hydrobiol. 1998;143(2):227-255. doi:10.1127/archiv-hydrobiol/143/1998/227

[47] Hunt GW, Chein SM. Seasonal distribution, composition and abundance of the planktonic Ciliata and Testacea of Cayuga Lake. Hydrobiologia. 1983;98(3):257-266. doi:10.1007/BF00021027

[48] Sonntag B, Posch T, Klammer S, Teubner K, Psenner R. Phagotrophic ciliates and flagellates in an oligotrophic, deep, alpine lake: contrasting variability with seasons and depths. Aquat Microb Ecol. 2006;43(2):193-207. doi:10.3354/ame043193

[49] Šimek K, Grujčić V, Nedoma J, et al. Microbial food webs in hypertrophic fishponds: Omnivorous ciliate taxa are major protistan bacterivores. Limnol Oceanogr. 2019;64(5):2295-2309. doi:10.1002/lno.11260

[50] Callieri C, Bertoni R, Contesini M, Bertoni F. Lake Level Fluctuations Boost Toxic Cyanobacterial "Oligotrophic Blooms." Dam HG, ed. PLoS One. 2014;9(10):e109526. doi:10.1371/journal.pone.0109526

[51] Capelli C, Ballot A, Cerasino L, Papini A, Salmaso N. Biogeography of bloom-forming microcystin producing and non-toxigenic populations of Dolichospermum lemmermannii (Cyanobacteria). Harmful Algae. 2017;67:1-12. doi:10.1016/j.hal.2017.05.004

[52] Stoeckerl DK, Evans2 GT. Effects of Protozoan Herbivory and Carnivory in a Microplankton Food Web. Vol 25.; 1985.

[53] Tarbe AL, Unrein F, Stenuite S, et al. Protist Herbivory: A Key Pathway in the Pelagic Food Web of Lake Tanganyika. Microb Ecol. 2011;62(2):314-323. doi:10.1007/s00248-011-9817-8

[54] Strom SL, Postel JR, Booth BC. Abundance, variability, and potential grazing impact of planktonic ciliates in the open subaratic Pacific Ocean. Prog Oceanogr. 1993;32(1-4):185-203. doi:10.1016/0079-6611(93)90013-4

[55] Yasindi AW, Taylor WD. Abundance, biomass and estimated production of planktonic ciliates in Lakes Victoria and Malawi. In: Aquatic Ecosystem Health and Management. Vol 6.; 2003:289-297. doi:10.1080/14634980301496

[56] Zingel P. Vertical and seasonal dynamics of planktonic ciliates in a strongly stratified hypertrophic lake. Hydrobiologia. 2005;547(1):163-174. doi:10.1007/s10750-005-4157-7

[57] Zingel P, Nõges T. Seasonal and annual population dynamics of ciliates in a shallow eutrophic lake. Fundam Appl Limnol. 2010;176(2):133-143. doi:10.1127/1863-9135/2010/0176-0133

[58] Dolan JR, Ritchie ME, Tunin-Ley A, Pizay MD. Dynamics of core and occasional species in the marine plankton: Tintinnid ciliates in the north-west Mediterranean Sea. J Biogeogr. 2009;36(5):887-895. doi:10.1111/j.1365-2699.2008.02046.x

[59] Karayanni H. Influence of ciliated protozoa and heterotrophic nanoflagellates on the fate of primary production in the northeast Atlantic Ocean. J Geophys Res. 2005;110(C7):C07S15. doi:10.1029/2004JC002602

[60] Olson MB, Strom SL. Phytoplankton growth, microzooplankton herbivory and community structure in the southeast Bering Sea: Insight into the formation and temporal persistence of an Emiliania huxleyi bloom. Deep Res Part II Top Stud Oceanogr. 2002;49(26): 5969-5990. doi:10.1016/S0967-0645(02)00329-6

[61] Dolan JR, Montagnes DJS, Agatha S, Coats DW, Stoecker DK. The Biology and Ecology of Tintinnid Ciliates.

(Dolan JR, Montagnes DJS, Agatha S, Coats DW, Stoecker DK, eds.). Chichester, UK: John Wiley & Sons, Ltd; 2012. doi:10.1002/9781118358092

[62] Hansen B, Bjornsen PK, Hansen PJ. The size ratio between planktonic predators and their prey. Limnol Oceanogr. 1994;39(2):395-403. doi:10.4319/lo.1994.39.2.0395

[63] Jakobsen HH, Hansen J. Prey Size Selection, Grazing and Growth Response of the Small Heterotrophic Dinoflagellate Gymnodinium Sp. and the Ciliate Balanion Comatum-a Comparative Study. Vol 158.; 1997.

[64] Johansson M. Annual variability in ciliate community structure, potential prey and predators in the open northern Baltic Sea proper. J Plankton Res. 2004;26(1):67-80. doi:10.1093/plankt/fbg115

[65] Ecol Prog Ser Per Jonsson MR. Particle Size Selection, Feeding Rates and Growth Dynamics of Marine Planktonic Oligotrichous Ciliates (Ciliophora: Oligotrichina) Marine ecology-progress series. Vol 27.; 1986.

[66] Mironova E, Telesh I, Skarlato S. Diversity and seasonality in structure of ciliate communities in the Neva Estuary (Baltic Sea). J Plankton Res. 2012;34(3):208-220. doi:10.1093/plankt/fbr095

[67] Petz W. Ciliate biodiversity in Antarctic and Arctic freshwater habitats - A bipolar comparison. In: European Journal of Protistology. Vol 39. Elsevier GmbH; 2003:491-494. doi:10.1078/0932-4739-00026

[68] Jürgens K, Skibbe O, Jeppesen E. Impact of metazooplankton on the composition and population dynamics of planktonic ciliates in a shallow, hypertrophic lake. Aquat Microb Ecol. 1999;17(1):61-75. doi:10.3354/ame017061

[69] Mieczan T, Adamczuk M. Ecology of Ciliates in Microbial Mats in Small Ponds: Relationship to Environmental Parameters (King George Island, Maritime Antarctica). Ann Zool Fennici. 2016;53(3-4):201-214. doi:10.5735/086.053.0409

[70] Yasindi AW, Taylor WD. The trophic position of planktonic ciliate populations in the food webs of some East African lakes. African J Aquat Sci. 2006;31(1):53-62. doi:10.2989/16085910609503871

[71] AGASILD H, ZINGEL P, KARUS K, KANGRO K, SALUJÕE J, NÕGES T. Does metazooplankton regulate the ciliate community in a shallow eutrophic lake? Freshw Biol. 2013;58(1):183-191. doi:10.1111/fwb.12049

[72] Kepner RL, Wharton RA, Coats DW. Ciliated protozoa of two antarctic lakes: Analysis by quantitative protargol staining and examination of artificial substrates. Polar Biol. 1999;21(5):285-294. doi:10.1007/s003000050364

[73] Pfister G, Arndt H. Food selectivity and feeding behaviour in omnivorous filter-feeding ciliates: A case study for Stylonychia. Eur J Protistol. 1998;34(4):446-457. doi:10.1016/S0932-4739(98)80013-8

[74] Weisse T. Functional diversity of aquatic ciliates. Eur J Protistol. 2017;61:331-358. doi:10.1016/j.ejop.2017.04.001

[75] Paffenhöfer GA, Sherr BF, Sherr EB. From small scales to the big picture: persistence mechanisms of planktonic grazers in the oligotrophic ocean. Mar Ecol. 2007;28(2):243-253. doi:10.1111/j.1439-0485.2007.00162.x

[76] Kisand V, Nõges T, Zingel P. Diel dynamics of bacterioplankton activity in eutrophic shallow Lake Võrtsjärv, Estonia. Hydrobiologia.

1998;380(1-3):93-102. doi:10.1023/A:1003444016726

[77] Macek M, Callieri C, Šimek K, Vázquez AL. Seasonal dynamics, composition and feeding patterns of ciliate assemblages in oligotrophic lakes covering a wide pH range. Arch fur Hydrobiol. 2006;166(2):261-287. doi:10.1127/0003-9136/2006/0166-0261

[78] Barka EA, Vatsa P, Sanchez L, et al. Taxonomy, Physiology, and Natural Products of Actinobacteria. Microbiol Mol Biol Rev. 2016;80(1):1-43. doi:10.1128/mmbr.00019-15

[79] Andrushchyshyn OP, Wilson KP, Williams DD. Ciliate communities in shallow groundwater: seasonal and spatial characteristics. Freshw Biol. 2007;52(9):1745-1761. doi:10.1111/j.1365-2427.2007.01806.x

[80] Ladau J, Eloe-Fadrosh EA. Spatial, Temporal, and Phylogenetic Scales of Microbial Ecology. Trends Microbiol. 2019;27(8):662-669. doi:10.1016/j.tim.2019.03.003

[81] Foissner W. Diversity and ecology of soil flagellates. In: Patterson, David J, Larsen J, ed. The Biology of Free-Living Heterotrophic Flagellates. Systematic. Clarendon Press Oxford; 1991: pp.93-112.

[82] Lynn DH. Ciliates. In: Schaeschter M, ed. Eukariotic Microbs. Academic Press; 2012:213-226.

[83] Jones DL, Clode PL, Kilburn MR, Stockdale EA, Murphy D V. Competition between plant and bacterial cells at the microscale regulates the dynamics of nitrogen acquisition in wheat (*Triticum aestivum*). New Phytol. 2013;200(3):796-807. doi:10.1111/nph.12405

[84] Gao Z, Karlsson I, Geisen S, Kowalchuk G, Jousset A. Protists: Puppet Masters of the Rhizosphere Microbiome. Trends Plant Sci. 2019;24(2):165-176. doi:10.1016/j.tplants.2018.10.011

[85] Bamforth SS. Water film fauna of microbiotic crusts of a warm desert. *J Arid Environ*. 2004;56(3):413-423. doi:10.1016/S0140-1963(03)00065-X

[86] Coûteaux M-M, Darbyshire J. Functional diversity amongst soil protozoa. *Appl Soil Ecol*. 1998;10(3):229-237. doi:10.1016/S0929-1393(98)00122-X

[87] Geisen S, Bonkowski M. Methodological advances to study the diversity of soil protists and their functioning in soil food webs. *Appl Soil Ecol*. June 2017. doi:10.1016/J.APSOIL.2017.05.021

[88] Buée M, de Boer W, Martin F, van Overbeek L, Jurkevitch E. The rhizosphere zoo: An overview of plant-associated communities of microorganisms, including phages, bacteria, archaea, and fungi, and of some of their structuring factors. *Plant Soil*. 2009;321(1-2):189-212. doi:10.1007/s11104-009-9991-3

[89] Bonfante P, Anca IA. Plants, mycorrhizal fungi, and bacteria: A network of interactions. *Annu Rev Microbiol*. 2009;63:363-383. doi:10.1146/annurev.micro.091208.073504

[90] Bamforth SS. Symposium on "Protozoan Ecology": The Role of Protozoa in Litters and Soils 1. *J Protozool*. 1985;32(3):404-409. doi:10.1111/j.1550-7408.1985.tb04035.x

[91] Bamforth SS. Interactions between protozoa and other organisms. *Agric Ecosyst Environ*. 1988;24(1-3):229-234. doi:10.1016/0167-8809(88)90068-0

[92] Krivtsov V, Liddell K, Bezginova T, et al. Ecological interactions of heterotrophic flagellates, ciliates and naked amoebae in forest litter of the

Dawyck Cryptogamic Sanctuary (Scotland, UK). *Eur J Protistol*. 2003;39(2):183-198. doi:10.1078/0932-4739-00883

[93] Petz W, Foissner W, Adam H. Culture, food selection and growth rate in the mycophagous ciliate Grossglockneria acuta Foissner, 1980: First evidence of autochthonous soil ciliates. *Soil Biol Biochem*. 1985;17(6):871-875. doi:10.1016/0038-0717(85)90149-X

[94] Foissner W. Soil protozoa as bioindicators: pros and cons, methods, diversity, representative examples. *Invertebr Biodivers as Bioindic Sustain Landscapes*. January 1999:95-112. doi:10.1016/B978-0-444-50019-9.50009-1

[95] Abraham JS, Sripoorna S, Dagar J, et al. Soil ciliates of the Indian Delhi Region: Their community characteristics with emphasis on their ecological implications as sensitive bio-indicators for soil quality. *Saudi J Biol Sci*. 2019;26(6):1305-1313. doi:10.1016/j.sjbs.2019.04.013

[96] Lara E, Berney C, Harms H, Chatzinotas A. Cultivation-independent analysis reveals a shift in ciliate 18S rRNA gene diversity in a polycyclic aromatic hydrocarbon-polluted soil. *FEMS Microbiol Ecol*. 2007;62(3):365-373. doi:10.1111/j.1574-6941.2007.00387.x

[97] Philippot L, Raaijmakers JM, Lemanceau P, Van Der Putten WH. Going back to the roots: The microbial ecology of the rhizosphere. *Nat Rev Microbiol*. 2013;11(11):789-799. doi:10.1038/nrmicro3109

[98] Geisen S. The bacterial-fungal energy channel concept challenged by enormous functional versatility of soil protists. *Soil Biol Biochem*. 2016;102:22-25. doi:10.1016/J.SOILBIO.2016.06.013

[99] Geisen S. The bacterial-fungal energy channel concept challenged by enormous functional versatility of soil protists. 2016. doi:10.1016/j.soilbio.2016.06.013

[100] Geisen S, Koller R, Hünninghaus M, Dumack K, Urich T, Bonkowski M. The soil food web revisited: Diverse and widespread mycophagous soil protists. *Soil Biol Biochem*. 2016;94:10-18. doi:10.1016/j.soilbio.2015.11.010

[101] Ekelund F. Enumeration and abundance of mycophagous protozoa in soil, with special emphasis on heterotrophic flagellates. *Soil Biol Biochem*. 1998;30(10-11):1343-1347. doi:10.1016/S0038-0717(97)00266-6

[102] Webster KE, Frost TM, Watras CJ, Swenson WA, Gonzalez M, Garrison PJ. Complex biological responses to the experimental acidification of Little Rock Lake, Wisconsin, USA. *Environ Pollut*. 1992;78(1-3):73-78. doi:10.1016/0269-7491(92)90012-Y

[103] Stoeck T, Edgcomb V. Role of Protists in Microbial Interactions with Hydrocarbons. In: *Handbook of Hydrocarbon and Lipid Microbiology*. Springer Berlin Heidelberg; 2010:2423-2434. doi:10.1007/978-3-540-77587-4_178

[104] Havens KE. Experimental Perturbation of a Freshwater Plankton Community: A Test of Hypotheses regarding the Effects of Stress. *Oikos*. 1994;69(1):147. doi:10.2307/3545295

[105] Jiang Y, Xu H, Hu X, Zhu M, Al-Rasheid KAS, Warren A. An approach to analyzing spatial patterns of planktonic ciliate communities for monitoring water quality in Jiaozhou Bay, northern China. *Mar Pollut Bull*. 2011;62(2):227-235. doi:10.1016/j.marpolbul.2010.11.008

[106] Caron DA, Sieburth JM. Response of Peritrichous Ciliates in Fouling Communities to Seawater-Accommodated Hydrocarbons. *Trans Am Microsc Soc*. 1981;100(2):183. doi:10.2307/3225801

[107] Parsons TR, Li WKW, Waters R. Some preliminary observations on the enhancement of phytoplankton growth by low levels of mineral hydrocarbons. *Hydrobiologia*. 1976;51(1):85-89. doi:10.1007/BF00007989

[108] Elloumi J, Carrias JF, Ayadi H, Sime-Ngando T, Boukhris M, Bouaïn A. Composition and distribution of planktonic ciliates from ponds of different salinity in the solar saltwork of Sfax, Tunisia. *Estuar Coast Shelf Sci*. 2006;67(1-2):21-29. doi:10.1016/j.ecss.2005.10.011

[109] Basuri CK, Pazhaniyappan E, Munnooru K, et al. Composition and distribution of planktonic ciliates with indications to water quality in a shallow hypersaline lagoon (Pulicat Lake, India). *Environ Sci Pollut Res*. 2020;27(15):18303-18316. doi:10.1007/s11356-020-08177-6

[110] Havens KE, Heath RT. Acid and aluminum effects on freshwater zooplankton: An in situ Mesocosm study. *Environ Pollut*. 1989;62(2-3):195-211. doi:10.1016/0269-7491(89)90187-5

[111] Wollmann K, Deneke R, Nixdorf B, Packroff G. Dynamics of planktonic food webs in three mining lakes across a pH gradient (pH 2-4). *Hydrobiologia*. 2000;433(1):3-14. doi:10.1023/A:1004060732467

[112] Peterson CH, Kennicutt II MC, Green RH, et al. Ecological consequences of environmental perturbations associated with offshore hydrocarbon production: a perspective on long-term exposures in the Gulf of Mexico. *Can J Fish Aquat Sci*. 1996;53(11):2637-2654. doi:10.1139/cjfas-53-11-2637

[113] Fleeger JW, Carman KR, Nisbet RM. Indirect effects of contaminants in aquatic ecosystems. *Sci Total Environ*. 2003;317(1-3):207-233. doi:10.1016/S0048-9697(03)00141-4

[114] Kachienga L, Jitendra K, Momba M. Metagenomic profiling for assessing microbial diversity and microbial adaptation to degradation of hydrocarbons in two South African petroleum-contaminated water aquifers. *Sci RepoRts |*. 2018;8:7564. doi:10.1038/s41598-018-25961-0

[115] Gertler C, Näther DJ, Gerdts G, Malpass MC, Golyshin PN. A mesocosm study of the changes in marine flagellate and ciliate communities in a crude oil bioremediation trial. *Microb Ecol*. 2010;60(1):180-191. doi:10.1007/s00248-010-9660-3

[116] Segovia BT, Lansac-Toha FM, de Meira BR, Cabral AF, Lansac-Tôha FA, Velho LFM. Anthropogenic disturbances influencing ciliate functional feeding groups in impacted tropical streams. *Environ Sci Pollut Res*. 2016;23(19):20003-20016. doi:10.1007/s11356-016-7185-0

[117] Moss JA, McCurry C, Tominack S, et al. Ciliated protists from the nepheloid layer and water column of sites affected by the Deepwater Horizon oil spill in the Northeastern Gulf of Mexico. *Deep Res Part I Oceanogr Res Pap*. 2015;106:85-96. doi:10.1016/j.dsr.2015.10.001

[118] Tirjaková E, Vďačný P. Recovery of ciliate communities from an oligotrophic mountain stream after a catastrophic wind damage. *Eur J Protistol*. 2013;49(4):526-537. doi:10.1016/j.ejop.2013.04.002

[119] Shabarova T, Salcher MM, Porcal P, et al. Recovery of freshwater microbial

communities after extreme rain events is mediated by cyclic succession. *Nat Microbiol*. January 2021:1-10. doi:10.1038/s41564-020-00852-1

[120] García CM, Niell FX. Seasonal change in a saline temporary lake (Fuente de Piedra, southern Spain). *Hydrobiologia*. 1993;267(1-3):211-223. doi:10.1007/BF00018803

[121] Delorenzo ME, Lauth J, Pennington PL, Scott GI, Ross PE. Atrazine effects on the microbial food web in tidal creek mesocosms. *Aquat Toxicol*. 1999;46(3-4):241-251. doi:10.1016/S0166-445X(98)00132-5

[122] Schindler DW. Experimental Perturbations of Whole Lakes as Tests of Hypotheses concerning Ecosystem Structure and Function. *Oikos*. 1990;57(1):25. doi:10.2307/3565733

[123] Rattan P, Ansari ZA, Sreepada RA. Ecology of a heterotrichous ciliate Fabrea salina from salterns of Bombay Coast, India. 1994. 35(2); 1994; 285-294.

[124] Nguyen-Viet H, Gilbert D, Mitchell EAD, Badot P-M, Bernard N. Effects of Experimental Lead Pollution on the Microbial Communities Associated with *Sphagnum fallax (Bryophyta)*. Microb Ecol. 2007;54, 232-241. doi.org/10.1007/s00248-006-9192-z

[125] Petz W, Foissner W. The effects of mancozeb and lindane on the soil microfauna of a spruce forest: A field study using a completely randomized block design. *Biol Fertil Soils*. 1989;7(3):225-231. doi:10.1007/BF00709653

[126] Jousset A, Lara E, Nikolausz M, Harms H, Chatzinotas A. Application of the denaturing gradient gel electrophoresis (DGGE) technique as an ef fi cient diagnostic tool for ciliate communities in soil. *Sci Total Environ*. 2010;408(5):1221-1225. doi:10.1016/j.scitotenv.2009.09.056

[127] Mondragón-Camarillo L, Zaragoza SR, Mendoza-López MR, Cabirol N, Macek M. Recovery of Soil Protozoan Community Structure Promoted by *M. sativa* After a Strong Pulse of Hydrocarbon Contamination. *Water Air Soil Pollut*. 2020;231(6):1-15. doi:10.1007/s11270-020-04618-7

Chapter 7

Ciliates as Symbionts

Rosaura Mayén-Estrada, Roberto Júnio Pedroso Dias, Mireya Ramírez-Ballesteros, Mariana Rossi, Margarita Reyes-Santos, Carlos Alberto Durán-Ramírez and Gerardo Cruz-Jiménez

Abstract

Although many ciliates are free-living, more than 140 families of ciliates (Alveolata, Ciliophora) include symbiotic species of animals. Symbiosis, defined as an interaction between two species, is analyzed in this chapter to show a wide diversity of symbiotic systems in ciliates (epibiosis, commensalism, mutualism, and parasitism), providing some data about ciliate strategies showing their success as symbionts. Some species are free-living as well symbionts, facultative symbionts, and obligate symbionts. Analysis of reconstructions of ancestral state evidence that the parasitism arose numerous times and independently among the lineages of ciliates. At least three evolutionary routes can be traced: (1) transition from free-living to mutualism and parasitism, (2) transition from free-living to parasitism, and (3) regression from parasitism to free-living. The evolution of the symbiosis in ciliates demonstrates a higher diversification rate concerning free-living ciliates. The analysis of the evolution of the life cycles complexity, exploring molecular data of the phases of the ciliate cycle in their hosts is also essential. We propose new approaches for an integrative study of symbiotic ciliates.

Keywords: Ciliophora, diversity, ecology, macroevolution, morphology, physiology, symbiosis, taxonomy

1. Introduction

Ciliates (Alveolata: Ciliophora) comprise free-living and symbiotic species. According to Corliss, [1] 2,600 species of ciliates have been described as symbionts, mainly of individuals of metazoan phyla. This is equivalent to 33% of all the known species of the phylum. They belong to eight classes (Armophorea, Heterotrichea, Litostomatea, Nassophorea, Oligohymenophorea, Plagiopylea, Phyllopharyngea and Spirotrichea), 31 orders, 151 families, and almost 700 genera [2]. These symbiotic ciliates have been reported in aerobic and anaerobic environments and from aquatic and terrestrial habitats [2, 3].

The term symbiosis can be defined as a sustained relationship between at least two individuals from different species, either living in direct contact or close enough to each other during a part or the whole life cycles of the partners. This interaction is transmitted vertically (from one generation to the next) or horizontally (acquired *de novo* in each generation). The intricate associations are believed to have an essential driving force in evolutionary biology, as a host and their symbiotic microbiota acclimatize on scales of short time [4].

Due to the diversity of symbioses, a classification system for symbiotic associations has been developed. This classification is based on several features: i) the dependence, where symbionts can be obligate or facultative; ii) specificity of the symbionts; iii) nutrients obtention, then biotrophic and necrotrophic symbionts are distinguished on the basis of whether nutrients are obtained from a living or dead partner, and iv) location of the symbionts, ectosymbionts or endosymbionts [5]. The symbiotic relationships can be categorized into mutualistic, commensalistic, or parasitic [2, 6]. The boundary between these categories sometimes is not clear, and there are frequent transitions between them.

Several papers have been focused on providing taxonomic reports for symbiotic ciliates, some of them as general works, and a few directed to certain groups [7–16], and some were focused on certain geographic areas [17–24]. Critical reviews of some species as *Balantidium coli* were done by Schuster and Ramirez-Avila [25]; for chonotrichs [26]; peritrichs [27] and suctorians [28].

Also, very different topics about ciliates and their hosts have been developed as shown: symbiotic interactions [epibiotic, hyperepibiotic, commensals, parasites (obligates and facultatives)], codiversification: [29–37]. Morphology (variation, molecular characterization): [38], clevellandellid, *Nyctotheroides*; [39], *Dicontophrya*; [40, 41] peritrichs. Taxonomy (new family, genus or species), redescription, revision: Apostomatia: [42]; Apostomatida: [43]; *Trichodina*: [44]; *Epistylis* and *Opercularia*: [45]; *Spirochona*: [46]; *Buetschlia* and *Charonina*: [31, 47–51]. Life cycles, encystment/excystment process: [52–54]. Pathogenicity, damages, infestation degree, virulence: [55–59]. Molecular and phylogeny: [30, 60–68]. Ecological aspects: [69, 70]. Immunity: [71, 72]. Stomatogenesis: [73]. Ultrastructure: [74].

Symbiotic systems between ciliates/animals are present in a broad spectrum of kingdom Animalia, and some examples are the following (animal group alphabetically arranged, different taxonomic levels): acari: [75]; amphipods: [76]; antilope: [77]; anuran: [78]; Asian elephant: [79]; baboon: [80]; bryozoans: [81]; buffaloes: [82]; capybara: [83–85]; cattle: [86]; chimpanzees: [87]; cirripedians: [88]; crustaceans: [89]; ctenophores: [90]; cuttlefish: [91]; dromedary camels: [92]; elephants: [93]; fishes: [94, 95]; frogs: [96]; great apes: [97]; horses: [98, 99]; humans: [100, 101]; polyps of hydras: [102]; insects: [103]; isopods: [104, 105]; kinorhynchs: [106]; llamas: [107]; maccacus: [108]; mammals: [109]; mollusks: [71, 76]; nematodes: [29, 110]; nemerteans: [13]; oligochaetes: [111, 112]; ostracods: [113]; polychaetes: [114, 115]; rhinoceroses: [116]; sea urchins: [117]; thoroughbreds: [118]; turbellarians: [119]; wood-feeding roaches: [120].

Some examples of ciliate taxa that include symbiotic species are the following:

Heterotrichea: Folliculinids attach to the integument of various invertebrates as bivalve shells, crustaceans exoskeleton, polychaete tubes, hydroid perisarcs, bryozoan tests, with a widespread occurrence [121], and may cause the skeletal eroding band or brown band diseases of scleractinian corals [2]; their life cycle includes a migratory swimming stage.

Spirotrichea: Hypotrichs are known mainly as free-living organisms, but some species such as *Euplotes balteatus* have been recorded in some sea urchins' intestinal tract [122]. Some species of stichotrichids as *Plagiotoma lumbrici* are endosymbionts of oligochaetes [123].

Armophorea: Class Armophorea includes clevellandellids as Nyctotheridae, with obligate endosymbionts usually as commensals of invertebrates and vertebrates; life cycles include a phase of the cyst [2].

Litostomatea: Trichostomes are symbionts of vertebrates as ruminants and foregut fermenters [2], including the human pathogen, *Balantidium coli*, species that have a life cycle including two phases: trophozoites and cysts [25]. This species has been considered to be included in a new genus, *Neobalantidium coli* [124]. The genus *Balantidium* has a more significant number of species that have been reported as endocommensals in the digestive tracts of a widely diverse range of metazoan, as mollusks, arthropods, fishes, reptiles, birds, and mammals [124]. In the rumen ecosystem, ciliates can account for up to 50% of the total microbial nitrogen, reaching densities of 10^5 to 10^6 cells/ml rumen fluid, being *Charonina ventriculi* one of the smallest rumen ciliates [125].

Ophryoscolecidae and Cycloposthiidae include species as endosymbionts of ruminants and equids, respectively [126]. Entodiniomorphid ciliates of the genus *Triplumaria* are found in the intestine of elephants and rhinoceroses [60]. Entodiniomorphida do not form cysts, and in non-ruminant mammals, the infections of hosts occur by coprophagy [47].

Phyllopharyngea: Chonotrichs live on marine and freshwater hosts and divide by forming external or internal buds [127], with a dimorphism where the adults live attached to several appendages of crustaceans, and the larva is free and swims to reach a new host [128].

Suctorians, as a rule, reproduce by different modes of budding, produce one to several larvae, with a short swimming existence, and then lose their cilia and metamorphose into adults or trophonts [127]. The non-ciliated mature stages of suctorians are usually sessile, attached to the substrate by a non-contractile stalk, and reproduce by ciliary larvae called swarmers or migrators [129].

Oligohymenophorea: Yi et al. [130] documented that the life cycle of *Ichthyophthirius multifiliis*, a parasite of fish, consists of three key developmental stages: the infective theront, the parasitic trophont, and the reproductive tomont.

Mesanophrys pugettensis, is a scuticociliate thata was observed with a diphasic life history, the larger phase or trophont, and the smaller phase resembling tomites [34], is a facultative parasite of the Dungeness crab. *Conchophthirus* species are generally considered an endocommensal inhabiting the mantle cavity of freshwater clams or mussels [30].

Thigmotrichids from several families were analyzed by Raabe [131–134], where species of Hemispeiridae are symbionts of the mantle cavity and nephridia of molluscan, those of Ancistrocomidae, Sphenopryidae and Thigmophryidae are ectosymbionts of mantle cavity and gills of molluscan, and Hysterocinetidae species were categorized as endoparasites of the gut of prosobranch mollusks; life cycles include tomites.

The apostomes is a small group of oligohymenophorean ciliates, with four major life histories: 1-exuviotrophic, that remain encysted on the exoskeleton of a crustacean host, and excyst to feed on exuvial fluid, reproducing during the host ecdysis, 2-sanguicolous, penetrate the cuticle of the host, feed on the cells and fluid of the hemocoel and reproduces, 3-chromidinid, found in the renal organs and opalinopsids found in the liver and intestines of cephalopods ingesting fluids and cells, 4-histotrophs, such as *Vampyrophrya* [135]. Apostome ciliates have life cycles typically involving crustaceans, with a non feeding microstome tomite and a macrostomous trophont [127]. Species of apostome of genus *Collinia* are endoparasites able to reproduce rapidly within the host that invariably kill the euphausiid within 40 hours of infection; *Gymnodinioides* genus includes exuviotrophic species that feed on the fluid within the exuviae of crustacean hosts and Landers *et al.*, [136] documented for *Gymnodinioides pacifica* the presence of trophonts, phoronts, tomonts and tomites. For *Synophrya* the phoront, hypertrophont, hypertomont, and hypertomites were observed [137].

Pilisuctorian ciliates spend most of their lives perched on cuticular setae of crustaceans, and complete their life cycle on a single host, having the stages tomite, tomont and trophont [138].

In peritrichs, a significant character is the scopula which is the region that originates the stalk to attach the organism to the substrate and modifies to a highly complicated adhesive apparatus in mobiline [127]; two phases are known, the trophont and the dispersive telotroch.

Species of sessile peritrichs genera such *Ambiphrya*, *Epistylis*, *Heteropolaria*, *Rhabdostyla*, and *Zoothamnium* are epibionts of zooplanktonic invertebrates, larval stages of aquatic insects, aquatic mollusks, crustaceans, fish, amphibians, and reptiles as the main groups of organisms [139]. Members of genus *Epistylis* have been reported as epibionts in several metazoans, but also as an important fish ectoparasite being considered an emerging pathogen [140]. Genus *Lagenophrys* comprises only symbiotic species of freshwater and marine crustaceans [89]. Trichodinids are the most devastating ectoparasites of cultured fish, causing severe damage [141], and for genus *Trichodina* about 300 species have been described, mostly from freshwater environments [142]. Also, there are reports of trichodinids from the gills of limpets [143] and have been documented as symbionts of a broad spectrum of aquatic and terrestrial invertebrates and vertebrates hosts [65]. *Trichodinella epizootica* is one of the most widely distributed freshwater trichodinids in Europe and Asia, but has also been reported from Africa, the Pacific region and North America [55]. *Urceolaria* includes species ectosymbionts of freshwater turbellarians, marine polychaetes, and mollusks; *Leiotrocha* species are ectocommensals and endocommensals of marine molluscans, and species of *Polycycla* are endocommensals of Holothuroidea [144].

2. Ecological relationships: Classical definitions and approaches

2.1 Epibiosis

Epibiosis is a facultative association of two organisms: the epibiont, which colonizes the surface of live substrates, and the basibiont, which hosts the epibionts [145]. Some species of epibiotic communities show preferences for specific location sites on the host [76]. According to Wahl and Mark [146], when the effects associated with epibiosis are neutral or positive for a basibiont species and beneficial for an epibiont species, selection should favor the evolution of the epibiotic relationship, which tends to increase specificity through evolutionary history. Although many epibiont ciliates are not harmful to their basibionts, some studies have shown that the epibionts can cause deleterious effects on their hosts [147–149].

Historically, studies involving epibiont ciliates focus on the following interests: new records and checklists [27, 28], descriptions of new taxa using morphological and molecular data [150], possible deleterious effects on hosts [149, 151], distribution and preferred sites of epibiont populations and communities [152], spatial and temporal distribution of the epibiotic relationship [153], laboratory rearing and experimentation studies [154–156], and even investigations into extrinsic and intrinsic factors involved in the kinetics of epibiont ciliate populations [157, 158].

2.2 Mutualism

Mutualism is a relationship with high metabolic dependence, where both organisms, ciliate and their hosts, obtain benefits [159, 160]. In the phylum Ciliophora, this type of relationship is seen, mainly in the subclass Trichostomatia, which includes the ciliates of the digestive tract of herbivorous mammals [161]. The symbiont ciliates represent approximately 2,600 of the described organisms, of which around 1000 species belong to the subclass Trichostomatia [2]. This subclass comprises ciliated protists, mostly mutualists of the digestive tract of several vertebrate hosts, with only one species showing parasitism in humans, *Balantidium coli* [2, 162, 163]. The subclass Trichostomatia is divided into three orders: Vestibuliferida, Entodiniomorphida, and Macropodiniida.

Ruminant ciliates and the host have a fundamental symbiosis relationship for the digestion and absorption of large amounts of plant material by the ruminant [164, 165]. On the one hand, the host provides an ideal environment for the survival of the symbiotic microbiota. The rumen is a strictly anaerobic environment, with temperatures ranging from 38 to 41° C, redox potential around 250 to 450 mV (millivolts), osmolarity ranging from 260 to 340 mOsm (millivolts), and pH levels between 5.0 and 7.5. Maintaining these characteristics is essential for microbial enzymatic activity to occur. In return, symbionts provide energy, protein, and vitamins to the host [166]. In energy terms, about 50–70% of the energy obtained by the host comes from the absorption of volatile fatty acids (VGAs) (eg. acetate, butyrate, and propionate), which are absorbed after the breakdown and fermentation of plant fiber by ruminal microorganisms [165]. Ciliates also represent a great source of protein for the ruminant (about 2 to 5%). Still, the ruminal microbiota also synthesizes B and K vitamins in sufficient quantities for the maintenance and growth of the animal. Due to the important participation in the physiology of the ruminant, the evolutionary dynamics of ruminal ciliates has been suggested as closely associated with the radiation of their hosts [167–169].

2.3 Commensalism and parasitism

Commensalism occurs when the symbiont inhabits in the host with no evident benefit or harm [170].

Parasitism, which is less common in ciliates, involves the parasites that usually cause disease being pathogens. They may be localized or spread throughout a host, defined as the independent and dominant member of the symbiotic pair. Here, the parasite inhabits on or inside the host to obtain resources and to harm it [171].

3. Ecological relationships: evolutionary approach

From an evolutionary point of view, there are species that are entirely free-living, those which can live equally well both free or as symbionts, species that are almost entirely symbiotic with only occasional periods of "free" existence during their life cycles (facultative symbionts), and species which are entirely symbiotic (obligate symbionts). Most of the well documented associations between Ciliophora and Metazoa are the ones leading to a certain degree of metabolic dependence. We will use in this topic the idea of metabolic dependence to define the ecological relationships: "free-living" (no metabolic dependence), "epibiont" (facultative metabolic dependence), "mutualistic" (mutual metabolic dependence) or "parasitic" (unilateral metabolic dependence, including commensalism).

For many years the evolutionary studies for Ciliophora were based only on morphological data, mainly those related to the ultrastructural characterization of its complex infraciliature [2]. However, in recent years this scenario has been modified with the implementation of modern tools that use multidisciplinary methods to integrate morphological, phylogenetic, molecular, and ecological data [161, 172–174]. A reliably dated phylogeny is fundamental to infer a broad macroevolutionary scenario for Ciliophora [172]. The inference of diversification rates from molecular phylogenies has increasingly been used to derive macroevolutionary patterns of lineages. Understanding how the different ecological relationships evolve in Ciliophora along time is a complex task that has been developed for many years. Different hypotheses about the origin and evolution of parasitic life have been proposed. Parasitologists suggest that the symbiotic way of life probably descended from free-living lineages that subsequently adapted to life in special habitats. Besides this, several authors suggest multiple origins of parasitism based on a comparison of morphological and ultrastructural aspects between them and their free life co-specifics [175], however, the processes that lead to its emergence are still imprecise [176–178].

Concerning the phylum Ciliophora, the vast majority of ciliates are categorized as free-living, and studies suggested that symbiosis apparently arising independently among various classes [179]. For genus *Tetrahymena* (subclass Hymenostomatia, order Hymenostomatida), all gradations of adaptations to symbiosis occur. There are species that live totally free, those that can live equally well both free and as symbionts, species that are almost entirely symbiotic with only occasional periods of "free" existence during their life cycles (optional symbionts), and species that are totally symbiotic (mandatory symbionts) [180]. Different transition routes between ecological associations have also been proposed, based on morphological and ecological characteristics. The first one proposes that free-living organisms assume habits of low metabolic dependence (mutualism, commensalism, among others), and with the strengthening of relationships, where they become parasites [176, 181]. The second hypothesis suggests that a free-living organism, when it comes into contact with a host accidentally, adapts itself to live both freely and within that host (optional parasite) [179], that is, free-living organisms adapt to live inside a host, which becomes something advantageous and increases fitness, making this a favorable way of life for the species.

Previous studies aimed to test these hypotheses based on phylogenetic analyzes of small groups within Ciliophora [174, 182, 183]. The macroevolutionary analyzes from the whole Ciliophora phylogeny presented **Figure 1** suggested that the ancestral way of life of the ciliates originated from a free-living organism and that the parasitic way of life arose numerous times and independently in Ciliophora, which was induced by two types of ancestors, free life and mutualistic (**Figure 1**). The transition to the parasitic way of life was recovered from two different origins: 1) a free-living ancestor evolved into a mutualistic organism and, later, to a parasitic organism, and 2) a free-living ancestor evolved into an organism parasite (highest number of cases). There are also cases where there has been a regression in the ciliate's way of life, where parasite clades evolved to free-living clades (**Figure 1**).

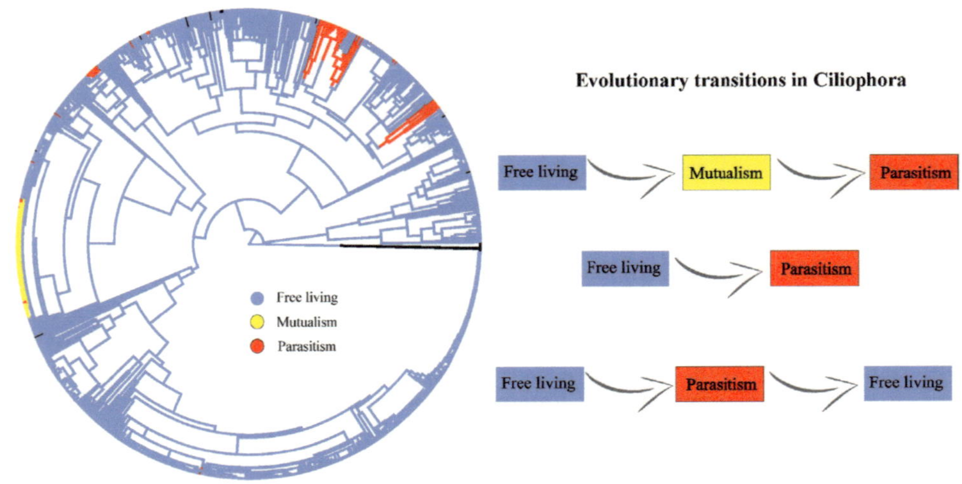

Figure 1.
Ancestral habit reconstruction for Ciliophora showing the main routes of transitions. Blue: Free-living. Yellow: Mutualism. Red: Parasitism/commensalism.

4. Future perspectives

The analytical improvement for morphological, ultrastructural, molecular, and evolutionary characterizations in Ciliophora culminated in an "Age of Integration", which several disciplines interact to infer patterns of biodiversity [184]. Although it is an age in full expansion, several gaps often prevent a study of diversity in its diverse areas in a complete way.

We are in a period of the paradigm shift, where Next Generation Sequencing (NGS) techniques have been applied exponentially, and, therefore, it is expected that new discoveries will emerge and new panoramas will be drawn on the diversity of the strains, as well as their respective ecological interactions. The transition from phylogenetic studies to phylogenomics is based on technological progress combined with exponential sequencing of molecular sequences (DNA, RNA), reduced associated costs, increased computational capacity, and improved analytical protocols. It is important to make efforts in studies to expand such technologies to lineages with little sampling in databases. For example, the classes Prostomatea, Oligohymenophorea, Litostomatea, and Phyllopharyngea, which present several examples of symbiosis, do not have available molecular sequences which prevents the evolutionary inferences of these lineages, requiring in the future more studies to refine the evolutionary hypotheses about the phylum. Efforts to expand metataxonomy using metagenomics and metatranscriptome methods have fed the databases exponentially in several lineages, revolutionized the analysis of environmental microbial diversity [175, 185, 186]. In fact, the generation of data for the target sequencing of phylogenetic, metagenomic, and metatranscriptomic markers is now reasonably well established, and several DNA sequencing platforms based on different technologies are currently available as well as different bioinformatics programs for each level of data extraction. However, due to the limited size of the molecular sequences produced by the platforms (~ 500 bp), phylogenetic estimates may be inadequate. With longer readings comes an improved phylogenetic signal, and we show that it is possible to employ a complete phylogenetic signal approach to taxonomically classify sequences and obtain a robust evolutionary structure of environmental diversity. New sequencing technologies such as nanopore sequencing, which offer long reads, improved the phylogenetic signal and more robust taxonomic patterns, can be an alternative in future studies [187].

With the significant increase in the number of available sequences from NGS sequencing, more effective and less subjective methodologies have been proposed to define the limits and number of independent evolutionary entities, to accelerate the biodiversity assessment process. In the last two decades, the field of species delimitation has intensified in relation to the number of methods available. For this, several methodologies have been proposed, based on biological [188], ecological [189], and molecular data [190], in addition to combining phylogenetic theory and population genetics [191–193]. The use of these methodologies in ciliates performed very recently to delimit organisms of free life, as species of the genus *Frontonia*, using the mitochondrial gene COX1 [194], species of the genus *Spirostomum*, using the ITS spacer region genes [195], and COI and 18S markers of the *Paramecium* genus.

Finally, several authors have emphasized the lack of studies on the distribution and occurrence of ciliates associated with Metazoa in natural conditions and the the lack of information on the ecology and interactions between epibionts and hosts. Few studies are exploring the natural history and complexity of life cycles, which makes it difficult to characterize optional and mandatory relationships. The absence of the characterization of the ciliate at the stage it is in the host, most studies, only in the environment, making it difficult to characterize the life cycle. Relevant information about habitat, life cycle, infection site is rare for Ciliophora [160, 196, 197].

Author details

Rosaura Mayén-Estrada[1*], Roberto Júnio Pedroso Dias[2],
Mireya Ramírez-Ballesteros[1], Mariana Rossi[3], Margarita Reyes-Santos[1†],
Carlos Alberto Durán-Ramírez[1] and Gerardo Cruz-Jiménez[1]

1 Laboratorio de Protozoología, Facultad de Ciencias, Universidad Nacional Autónoma de México, Mexico City, Mexico

2 Laboratório de Protozoologia, Programa de Pós-graduação em Biodiversidade e Conservação da Natureza, Universidade Federal de Juiz de Fora, Juiz de Fora, Brazil

3 Laboratório de Protozoologia, Instituto de Ciências Biológicas, Universidade Federal de Juiz de Fora, Juiz de Fora, Brazil

*Address all correspondence to: rme2@ciencias.unam.mx

† Deceased.

References

[1] Corliss JO. Biodiversity, classification, and numbers of species of protists. Nature and human society. The quest for a sustainable world. National Academy Press, Washington. 2000;130-155

[2] Lynn DH. The Ciliated Protozoa. Characterization, classification, and guide to the literature. Ontario: Springer; 2008. 605 p

[3] Rosati G. Ectosymbiosis in Ciliated Protozoa. In: Seckbach J, editor. Symbiosis. Mechanisms and Model Systems Cellular Origin, Life in Extreme Habitats and Astrobiology. Dordrecht: Springer; 2004. p. 475-488

[4] Oulhen N, Schulz BJ, Carrier TJ. English translation of Heinrich Anton de Bary's 1878 speech,'Die Erscheinung der Symbiose' ('De la symbiose'). 2016;69: 131-139

[5] Paracer S, Ahmadjian V. Symbiosis: An Introduction to Biological Associations. New York: Oxford University Press; 2000. 291 p

[6] Dziallas C, Allgaier M, Monaghan MT, Grossart HP. Act together—implications of symbioses in aquatic ciliates. Frontiers in Microbiology. 2012;3(288):1-17

[7] Nenninger U. Die Peritrichen der Umgebung von Erlangen mit besonderer Berüksichtigung ihrer Wirtsspezifität. Zoologische Jahrbücher (Systematik). 1948;77:169-266.

[8] Wolska M. Study on the family Blepharocorythidae Hsiung. I. Preliminary remarks. Acta Protozoologica. 1966;4(12):97-103

[9] Sprague V, Couch JA. An annotated list of protozoan parasites, hyperparasites, and commensals of decapod crustaceans. Journal of Protozoology. 1971;18:526-537

[10] Morado JM, Small EB. Ciliate parasites and related diseases of Crustacea: a review. Reviews in Fisheries Science. 1995;3(4):275-354

[11] Bojko J, Ovcharenko M. Pathogens and other symbionts on the Amphipoda: taxonomic diversity and pathological significance. Diseases of Aquatic Organisms. 2019;136:3-36

[12] Clamp JC, Chatterjee T, Fernandez-Leborans G. Occurrence of *Epistylis anastatica* (Linnaeus, 1767) (Ciliophora: Peritrichia) on *Mesocyclops isabellae* Dussart & Fernando (Crustacea: Copepoda: Cyclopoida) in India, with an annotated checklist of species of *Epistylis* reported as epibionts of cyclopoid copepods and resources for their identification. Zootaxa. 2016;4168(3):525-540

[13] McDermott JJ. Nemerteans as hosts for symbionts: a review. Journal of Natural History. 2006;40(15-16): 1007-1020

[14] Edgerton BF, Evans LH, Stephens FJ, Overstreet RM. Synopsis of freshwater crayfish diseases and commensal organisms. Aquaculture. 2002;296(1-2): 57-135

[15] Matthes D, Guhl W, Haider G. Suctoria und Urceolariidae. Protozoenfauna Band 7/1. Stutgart: Gustav Fischer Verlag; 1988. 309p

[16] de Jager GP, Basson L. Taxonomic assessment of three North American trichodinids by re-evaluating the taxon validity of *Trichodina heterodentata* Duncan, 1977 (Peritrichia). Acta Protozoologica. 2019;58:125-139

[17] Schödel H. A synopsis of the limnetic epizoic Peritrichia (Ciliophora) on invertebrates in Western-, Middle- and Eastern Europe. Denisia. 2018;41:47-294

[18] Chatterjee T, Fernandez-Leborans G, Ramteke D, Ingole B.

New records of epibiont ciliates (Ciliophora) from Indian coast with descriptions of six new species. Cahiers de Biologie Marine. 2013;54:143-159

[19] Chatterjee T, Dovgal IV, Nanajkar M, Fernandes V. Report of epibiont ciliates (Ciliophora: Suctorea) on pelagic ostracods (Crustacea: Ostracoda) from the Arabian Sea. Zootaxa. 2019a;4695(4):368-384

[20] Purushothaman A, Dovgal I, Francis SV, Padmakumar KB. Observation of a suctorian ciliate *Ephelota coronata* on the calanoid copepod *Pontella spinipes* in the southeastern Arabian Sea. Symbiosis. 2020;81(3):321-327

[21] Fernandez-Leborans G, Cárdenas CA. Epibiotic protozoan communities on juvenile southern king crabs (*Lithodes santolla*) from subantartic areas. Polar Biology. 2009;32:1693-1703

[22] Viljoen S, van As JG. Sessile peritrichs (Ciliophora: Peritricha) from fresh-water fish in the Transvaal, South Africa. South African Journal of Zoology. 1985;20(3):79-96

[23] O'Donoghue PJ, Adlard RD. Catalogue of protozoan parasites recorded in Australia. Memoirs of the Queensland Museum. 2000;45(1):1-164

[24] Dovgal I, Chatterjee T, Subba Rao DV, Chan BKK, De Troch M. New records of *Praethecacineta halacari* (Schulz) (Suctorea: Ciliophora) from Taiwan, Tanzania and Canada. Marine Biodiversity Records. 2009;2:e136, 1-3

[25] Schuster FL, Ramirez-Avila L. Current world status of *Balantidium coli*. Clinical Microbiology Reviews. 2008;21:626-638

[26] Fernandez-Leborans G. A review of the species of protozoan epibionts on crustaceans. III. Chonotrich ciliates. Crustaceana. 2001;74(6):581-607

[27] Fernandez-Leborans G, Tato-Porto ML. A review of the species of protozoan epibionts of crustaceans. I. Peritrich ciliates. Crustaceana. 2000a; 73:643-683

[28] Fernandez-Leborans G, Tato-Porto ML. A review of the species of protozoan epibionts of crustaceans. II. Suctorian ciliates. Crustaceana. 2000b;73(10):1205-1237

[29] Baldrighi E, Dovgal I, Zeppilli D, Abibulaeva A, Michelet C, Michaud E, Franzo A, Grassi E, Cesaroni L, Guidi L, Balsamo M, Sandulli, R, Semprucci, F. The cost for biodiversity: records of ciliate-nematode epibiosis with the description of three new suctorian species. Diversity. 2020;12:(224)1-25.

[30] Antipa GA, Strüder-Kypke MC, Lynn DH. Molecular phylogeny, taxonomic relationships and North American distribution of *Conchophthirus* (Conchophthiridae, Scuticociliatia). Aquatic Ecosystem Health and Management. 2020;23(1):58-68.

[31] Li M, Wang J, Zhu D, Gu Z, Zhang J, Gong X. Study of *Apiosoma piscicola* (Blanchard 1885) occurring on fry of freshwater fishes in Hongze, China with consideration of the genus *Apiosoma*. Parasitology Research. 2008;102:931-937

[32] Kozloff EN, Norenberg JP. *Raabella concinna* sp. nov. and *Hypocomina obstipa* sp. nov., ciliates of the family Ancistrocomidae parasitizing marine mollusks on the Pacific coast of North America. Cahiers de Biologie Marine. 2013;54:175-180

[33] Xu K, Lei, Y, Al-Rasheid KAS, Song W. Two new ectoparasitic ciliates, *Sphenophrya solinis* sp. nov. and *Planeticovorticella paradoxa* sp. nov. (Protozoa: Ciliophora), from marine molluscs. Journal of the Marine Biological Association of the United Kingdom. 2011;91(2):265-274

[34] Morado JM, Small EB. Morphology and stomatogenesis of *Mesanophrys pugettensis* n. sp. (Scuticociliatida: Orchitophryidae), a facultative parasite ciliate of the dungeness crab, *Cancer magister* (Crustacea: Decapoda). Transactions of the American micro-scopical Society. 1994;113(3):343-364

[35] Fernandez-Leborans G, Dávila P, Cerezo E, Contreras C. Epibiosis and hyperepibiosis on *Pagurus bernhardus* (Crustacea: Decapoda) from the west coast of Scotland. Journal of the Marine Biological Association of the United Kingdom. 2013;93(5):1351-1362

[36] Van As JG, Viljoen S.A. Taxonomic study of sessile peritrichs (Ciliophora: Peritricha) associated with crustacean fish ectoparasites in South Africa. South African Journal of Zoology. 1984;19(4):275-279

[37] Lynn DH, Gómez-Gutiérrez J, Strüder-Kypke MC, Shaw CT. Ciliate species diversity and host-parasitoid codiversification in *Pseudocollinia* infecting krill, with description of *Pseudocollinia similis* sp. nov. Diseases of Aquatic Organisms. 2014;112:89-102

[38] Li M, Sun ZY, Grim JN, Ponce-Gordo F, Wang G, Zou H, Li W, Wu S. Morphology of *Nyctotheroides hubeiensis* Li et al. 1998 from frog hosts with molecular phylogenetic study of clevelandellid ciliates (Armophorea, Clevelandellida). Journal of Eukaryotic Microbiology. 2016;63(6):751-759

[39] Fokam Z, Nana PA, Ngassam P, Bricheux G, Bouchard P, Viguès B, Sime-Ngando T. Morphological characterization of two new species of *Dicontophrya* (Ciliophora: Astomatia: Contophryidae) comensal of earthworms (Oligochaeta: Annelida) of Edeba and Nkolbikogo (Cameroon). International Journal of Biosciences. 2015b;7(1):49-57

[40] Zhou T, Wang Z, Yang H, Gu Z. Morphological and molecular identification of epibiontic sessilid *Epistylis semiciculus* n. sp. (Ciliophora, Peritrichia) from *Procambarus clarki* (Crustacea, Decapoda) in China. LJP: Parasites and Wildlife. 2019a;10:289-298

[41] Clamp JC. The occurrence of *Lagenophrys aselli* (Ciliophora: Peritricha: Lagenophryidae) in North America and a description of environmentally-induced morphological variation in th especies. Transactions of the American microscopical Society. 1988;107(1):17-27

[42] Chantangsi C, Lynn DH, Rueckert S, Prokopowicz AJ, Panha S, Lenader BS. *Fusiforma themisticola* n. gen. n. sp., a new genus and species of apostome ciliate infecting the hyperiid amphipod *Themisto libellula* in the Canadian Beaufort Sea (Arctic Ocean), and establishment of the Pseudocolliniidae (Ciliophora, Apostomatia). Protist. 2013;164:793-810

[43] Gómez-Gutiérrez J, Strüder-Kypke MC, Lynn DH, Shaw TC, Aguilar-Méndez MJ, López-Cortés A, Martínez-Gómez S, Robinson CJ. *Pseudocollinia brintoni* gen. nov., sp. nov. (Apostomatida: Colliniidae), a parasitoid ciliate infecting the euphausiid *Nyctiphanes simplex*. Diseases of Aquatic Organisms. 2012;99:57-78

[44] Pinto HA, Wieloch AH, de Melo AL. Uma nova espécie de *Trichodina* Ehrenberg, 1838 (Ciliophora: Trichodinidae) em *Biomphalaria schrammi* (Crosse, 1864) (Mollusca: Planorbidae). Lundiana. 2006;7(2):121-124.

[45] Zhou T, Wang Z, Yang H, Gu Z. Two new colonial peritrich ciliates (Ciliophora, Peritrichia, Sessilida) from China: with a note on taxonomic distinction between Epistylididae and Operculariidae. European Journal of Protistology. 2019b;70:17-31

[46] Dovgal IV, Grigorovich IA. *Spirochona americana* sp. n. (Ciliophora,

Chonotrichia)- a new North American freshwater chonotrich species from the amphipod (*Gammarus lacustris*). Vestnik Zoologii. 2001;35(5):77-80

[47] Gürelli G. New entodiniomorphid ciliates, *Buetschlia minuta* n. sp., *B. cirrata* n. sp., *Charonina elephanti* n. sp., from Asian elephants of Turkey. Zootaxa. 2019;4545(3):419-433

[48] Grim JN, Buonanno F. A re-description of the ciliate genus ans type species, *Balantidium entozoon*. European Journal of Protistology. 2009;45:174-182

[49] van Hoven W, Gilchrist FMC, Hamilton-Attwell VL. A new family, genus, and seven new species of Entodiniomorphida (Protozoa) from the gut of African rhinoceros. Journal of Protozoology. 1988;35(1):92-97

[50] Bradbury PC, Clamp JC. *Hyalophysa lwoffi* sp. n. from the freswater shrimp *Palaemonetes paludosus* and revision of the genus *Hyalophysa*. Journal of Protozoology. 1973;20(2):210-213

[51] Van As JG, Basson L. A further contribution to the taxonomy of the Trichodinidae (Ciliophora: Peritrichia) and a review of the taxonomic status of some fish ectoparasitic trichodinids. Systematic Parasitology. 1989;14:157-179

[52] Bassham RD, Browning JS, Landers SC. The complete life cycle of the inusual apostome *Hyalophysa clampi* (Ciliophora, Apostomatida), a symbiont of crayfish in Alabama (USA). European Journal of Protistology. 2020;72:125654

[53] Lucas, C. A study of excystation in *Nyctotherus ovalis* with notes on other intestinal protozoa of the cockroach. The Journal of Parasitology. 1928;14(3): 161-175

[54] Segade P, García N, García-Estevez JM, Arias C, Iglesias R. Encystment/excystment response and serotypic variation in the gastropod parasite *Tetrahymena rostrata* (Ciliophora, Tetrahymenidae). Parasitology Research. 2016;115:771-777

[55] Basson, L. First records of trichodinid ectoparasites (Ciliophora: Peritrichia) from introduced freshwater fishes in Tasmania, Australia, with comment son pathogenicity. Acta Protozoologica. 2010;49:253-265

[56] Small HJ, Neil DM, Taylor AC, Bateman K, Coombs GH. A parasitic scuticociliate infection in the Norway lobster (*Nephrops norvegicus*). Journal of Invertebrate Pathology. 2005;90:108-117

[57] Xie P, Sanderson B, Frost T, Magnuson JJ. Manipulation of host density affects infestation of a peritrich ciliate (*Epistylis lacustris*) on a calanoid copepod (*Leptodiaptomus minutus*) in Crystal lake, Wisconsin, USA. Journal of Freshwater Ecology. 2001;16(4):557-564

[58] Hanamura Y, Nagasaki K. Occurrence of the sandy beach mysids *Archaeomysis* spp. (Mysidacea) infested by epibiontic peritrich ciliates (Protozoa). Crustacean Research. 1996;25:25-33

[59] Swennes AG, Noe JG, Craig Findly R, Dickerson HW. Differences in virulence between two serotypes of *Ichthyophthirius multifiliis*. Diseases of Aquatic Organisms. 2006;69:227-232

[60] Ito A, Honma H, Gürelli G, Cocmen B, Mishima T, Nakai Y, Imai S. Redescription of *Triplumaria selenica* Latteur et al., 1970 (Ciliophora, Entodiniomorphida) and its phylogenetic position base don the infraciliary bands an 18SSU rRNA gene sequence. European Journal of Protistology. 2010;46:180-188

[61] Ito A, Ishihara M, Imai S. *Bozasella gracilis* n. sp. (Ciliophora, Entodiniomorphia) from Asian elephant and phylogenetic analysis of entodiniomorphids and vestibuliferids.

European Journal of Protistology. 2014;50:134-152

[62] Lu B, Shen Z, Zhang Q, Hu X, Warren A, Song W. Morphology and molecular analyses of four epibiotic peritrichs on crustacean and polychaete hosts, including descriptions of two new species (Ciliophora, Peritrichia). European Journal of Protistology. 2020;73:125670

[63] Snelling T, Pinloche E, Worgan HJ, Newbold CJ, McEwan NR. Molecular phylogeny of *Spirodinium equi*, *Triadinium caudatum* and *Blepharocorys* sp. from the equine hindgut. Acta Protozoologica. 2011;50:319-326

[64] Tang FH, Zhao YJ, Warren A. Phylogenetic analyses of trichodinids (Ciliophora, Oligohymenophorea) inferred from 18S rRNA gene sequence data. Current Microbiology. 2013;66: 306-313

[65] Xu K, Lei Y, Song W, Choi JK, Warren A. Diagnoses and probable phylogenetic relationships of the genera in the family Trichodinidae (Ciliophora, Peritrichia). The Yellow Sea. 2000;6:42-49

[66] Gao F, Gao S, Wang P, Katz L, Song W. Phylogenetic analyses of cyclidiis (Protista, Ciliophora, Scuticociliatia) based on multiple genes suggest their close relationship with thigmotrichichids. Molecular Phylogenetics and Evolution. 2014;75:219-226.

[67] Affa'a FM, Hickey DA, Strüder-Kypke M, Lynn DH. Phylogenetic position of species in the genera *Anoplophrya*, *Plagiotoma*, and *Nyctotheroides* (Phylum Ciliophora), endosymbiotic ciliates of annelids and anurans. Journal of Eukaryotic Microbiology. 2004;51(3):301-306

[68] Clamp JC, Bradbury PC, Strüder-Kypke MC, Lynn DH. Phylogenetic position of the apostome ciliates (Phylum Ciliophora, Subclass Apostomatia) tested using small subunit rRNA gene sequences. Denisia. 2008;23:395-402

[69] Fokam Z, Nana PA, Moche K, Bricheux G, Bouchard P, Ngassam P, Sime-Ngando T. Influence of soil physicochemical parameters on the abundance of *Paracoelophrya polymorphus* (Ciliophora; Radiophryidae) commensal of earthworms (Annelida: Glossoscolecidae) collected in Bambui (North-West Cameroon). Journal of Biodiversity and Environmental Sciences. 2015a;6(2): 376-389

[70] Ogut H, Akyol A, Alkan MZ. Seasonality of *Ichthyophthirius multifiliis* in the trout (*Oncorhynchus mykiss*) farms of the Eastern Black Sea region of Turkey. Turkish Journal of Fisheries and Aquatic Sciences. 2005;5:23-27

[71] Xu K, Song W. Two trichodinid ectoparasites from marine molluscs in the Yellow Sea, off China, with the description of *Trichodina caecellae* n sp. (Protozoa: Ciliophora: Peritrichia). Systematic Parasitology. 2008;69:1-11

[72] Heinecke RD, Buchmann K. Inflammatory response of rainbow trout *Oncorhynchus mykiss* (Walbaum, 1792) larvae against *Ichthyophthirius multifiliis*. Fish & Shelfish Immunology. 2013;34: 521-528

[73] Njine T, Ngassam P. Morphogenesis of bipartition in two species of the genus *Ptychostomum* (Ciliophora, Hysterocinetidae). European Journal of Protistology. 1993;29:396-406

[74] Landers SC. Morphogenesis in *Hyalophysa chattoni* (Ciliophora: Apostomatida): reduction of oral polykineties in the encysted phoront. Archiv für Protistenkunde. 1997;148: 389-397

[75] Dovgal I, Chatterjee T, Ingole B. An overview of suctorian ciliates (Ciliophora, Suctorea) as epibionts of

halacarid mites (Acari, Halacaridae). Zootaxa. 2008a;1810:60-68

[76] Fenchel T. On the ciliate fauna associated with the marine species of the amphipod genus *Gammarus* J. G. Fabricius. Ophelia. 1965;2(2):281-303

[77] Latteur B. *Diplodinium archon* n. sp., ciliate ophryoscolescide du rumen de l'antilope *Tragelaphus scriptus* Pallas. Annales de la Société Belge de Médecine Tropicale. 1966;46(6):727-740

[78] Affa'a FM. Observations morphologiques sur deux nyctothéres (ciliés hétérotriches) commensaux de batraciens anoures du Québec. Canadian Journal of Zoology. 1991;69:2765-2770

[79] Gürelli G. Intestinal ciliate fauna of the Asian elephant from Gaziantep, Turkey and the description of *Brevitentaculum antebum* n. g., n. sp. European Journal of Protistology. 2016;56:26-31

[80] Al-Tayib O, Abdoun KA. *Balantidium coli* infection in hamadryas baboon (*Papio hamadryas*) in Saudi Arabia: a case report. The Journal of Animal and Plant Sciences. 2013;23(3):939-943.

[81] Chatterjee T, Dovgal IV. A checklist of ciliate epibionts (Ciliophora) found on bryozoans. Zootaxa. 2020;4896(4): 547-559

[82] Roy BC, Mondal MMH, Talukder MH, Majumder S. Prevalence of *Balantidium coli* in buffaloes at different areas of Mymensingh. Journal of the Bangladesh Agricultural University. 2011;9(1):67-72

[83] Martinele I, Detoni ML, Rust NM, D'Agosto M. Protozoários ciliados (Protista, Ciliophora) no conteúdo do rúmen e do retículo de bovinos. Revista Brasileira de Zoociências. 2007;9(1):63-66

[84] Hollande A, Batisse A. Contribution à l'étude des infusoires parasites du coecum de l'hydrocheire (*Hydrocheirus capybara* L.). I. La famille des Cycloposthiidae. Memórias do Instituto Oswaldo Cruz. 1959;57(1):1-16

[85] Ito A, Imai S. Ciliates from the cecum of capybara (*Hydrochoerus hydrochaeris*) in Bolivia. 2. The family Cycloposthiidae. European Journal of Protistology. 2000;36:169-200

[86] Gürelli G, Akman FTB. Rumen ciliate biota of domestic cattle (*Bos taurus taurus*) in Istanbul, Turkey and infraciliature of *Metadinium medium* (Entodiniomorphida, Ophryoscolecidae). Acta Protozoologica. 2017;56:171-180

[87] Pomajbíková K, Petrzelková KJ, Petrásová J, Profousová I, Kalousová B, Jirjú M, Sá RM, Modrý D. Distribution of the entodiniomorphid ciliate *Troglocorys cava* Tokiwa, Modrý, Ito, Pomajbíková, Petrzelková & Imai, 2010, (Entodiniomorphida: Blepharocorythidae) in wild and captive chimpanzees. Journal of Eukaryotic Microbiology. 2012;59(1):97-99

[88] Arvy L, Batisse A, Lacombe D. Peritriches épizoïques dans la chambre branchiale des Balanidae (Crustacea: Cirripedia) *Epistylis nigrellii* n. sp. *E. horizontalis* (Chatton, 1930). Annales de Parasitologie Humaine et Comparee. 1969;44(4):351-374

[89] Clamp JC. Revision of the family Lagenophryidae Bütschli, 1889 and description of the family Usconophryidae n. fam. (Ciliophora; Peritricha). Journal of Protozoology. 1991;38(4):355-377

[90] Estes AM, Reynolds BS, Moss AG. *Trichodina ctenophorii* n. sp., a novel symbiont of ctenophores of the Northern coast of the Gulf of Mexico. Journal of Eukaryotic Microbiology. 1997;44(5):420-426.

[91] Ma M, Lu B, Fan X, Shi Y, Chen X. Taxonomic clarification of a well-known pathogenic scuticociliate, *Miamensis*

avidus Thompson & Moewus, 1964 (Ciliophora, Scuticociliatia). Journal Ocean University China (Oceanic and Coastal Sea Research). 2018;17(6):1-12

[92] Gürelli G, Daw AFOE. Endosymbiotic ciliated protozoan biota of dromedary camels and domestic cattle in Tunisia. Zootaxa. 2020;4859(3):409-418

[93] Gürelli G, Ito A. Intestinal ciliated protozoa of the Asian elephant *Elephas maximus* Linnaeus, 1758 with the description of *Triplumaria izmirae* n. sp. European Journal of Protistology. 2014;50:25-32

[94] Corrêa LL, Ceccarelli PS, Tavares-Dias M. An outbreak of *Ichthyophthirius multifiliis* (Ciliophora: Ichthyophthiriidae) in wild endemic fish fauna *Steindachneridium parahybae* (Siluriformes: Pimelodiae) in Brazil. Annals of Parasitology. 2019;65(4): 417-421

[95] El-Tantawy SAM, Reda ESA, Abdel-Aziz A, Abou El-Nour MF, Rady I. *Apiosoma* spp. and *Scopulata epibranchialis* infesting Nile perch fish *Lates niloticus* in Dakahlia Province, Egypt. New York Science Journal. 2013;6(6):111-118

[96] Pritchett KR, Sanders GE. Epistylididae ectoparasites in a colony of African clawed frogs (*Xenopus laevis*). Journal of the American Association for Laboratory Animal Science. 2007;46(2): 86-91

[97] Pomajbíková K, Petržlková KJ, Profousová I, Petrášová J, Modrý D. Discrepancies in the occurrence of *Balantidium coli* between wild and captive African great apes. Journal of Parasitology. 2010;96(6):1139-1144

[98] Gürelli G, Göçmen B. Intestinal ciliate composition found in the feces of the Turk rahvan horse *Equus caballus* Linnaeus, 1758. European Journal of Protistology. 2011;47:245-255

[99] Gürelli G, Canbulat S, Aldayarov N. Fecal ciliate composition of domestic horses (*Equus caballus* Linnaeus, 1758) living in Kyrgyzstan. Zootaxa. 2015; 4039(1):145-156

[100] Bel GS, Couret M. *Balantidium coli* infection in man. The Journal of Infectious Diseases. 1910;7(5):609-624

[101] Bellanger AP, Scherer E, Cazorla A, Grenouillet F. Dysenteric syndrome due to *Balantidium coli*: a case report. New Microbiologica. 2013;36:203-205

[102] Warren A, Robson EA. The identity and occurrence of *Kerona pediculus* (Ciliophora: Hypotrichida), a well-known epizoite of *Hydra vulgaris* (Cnidaria: Hydrozoa). Zoologische Verhandelingen. 1998;323(31);235-245

[103] Tanahashi M, Meng XY, Fukatsu T. A novel symbiotic ciliate (Ciliophora: Peritrichia) in the hindgut of a stag beetle (Coleoptera: Lucanidae). Zoological Science. 2017;34(3):217-222

[104] Santha Kumari V, Nair NB. Some ciliates from the marine wood-boring isopod *Sphaeroma*. Indian Journal of Fisheries. 1985;32(2):215-223

[105] Jankowski AW. *Eufolliculina brunea* sp. nov. (Ciliophora, Heterotrichida), a symbiont of wood-boring *Limnoria* isopods. Zoosystematica Rossica. 2009b;18(2):179-183

[106] Dovgal I, Chatterjee T, Ingole B, Nanajkar. First report of *Limnoricus ponticus* Dovgal & Lozowski (Ciliophora: Suctorea) as epibionts on *Pycnophyes* (Kinorhynch) from the Indian Ocean with keys to species of the genus *Limnoricus*. Cahiers de Biologie Marine. 2008b;49:381-385

[107] Cerón-Cuchi M, Marcoppido G, Dekker A, Fondevila M, De la Fuente G, Morici G, Cravero S. Ciliate protozoa of the forestomach of llamas (*Lama glama*) from locations at different altitude in Argentina. Zootaxa. 2016;4067(1):49-56

[108] Hinde K. Milk composition varies in relation to the presence and abundance of *Balantidium coli* in the mother in cautive *Rhesus macaques* (*Macaca mulatta*). American Journal of Primatology. 2007;69:625-634

[109] Nakauchi K. The prevalence of *Balantidium coli* infection in fifty-six mammalian species. Journal of Veterinary Medical Science. 1999;61(1):63-65

[110] Chatterjee T, Dovgal I, Fernandez-Leborans G. A checklist of suctorian epibiont ciliates (Ciliophora) found on meiobenthic marine nematodes. Journal of Natural History. 2019b;53(33-34):2133-2143

[111] Fokam Z, Ngassam P, Boutin CL, Zebazé-Togouet SH. Trois espèces nouvelles de *Coelophrya*, ciliés astomes endocommensaux d'*Alma nilotica* (oligochète terricole) du Cameroun. Bulletin de la Société d'histoire Naturelle de Toulouse. 2008;144:27-33

[112] Ngassam P, Grain J. Description of the new genus *Hysterocinetoides* (Ciliophora, Hysterocinetidae) and revision of the systematics of the subclass Hysterocinetia. Protistology. 2002;38:11-17

[113] Jankowski AW. *Tokophrya sibirica* sp. nov. (Ciliophora: Acinetida), a symbiont of freshwater ostracods inIrkutsk Province. Zoosystematica Rossica. 2014;23(2):174-177

[114] Lom J, Kozloff EN. The ultrastructure of *Phalacrocleptes verruciformis*, an unciliated ciliate parasitizing the polychaete *Schizobranchia insignis*. The Journal of Cell Biology. 1967;33:355-364

[115] Jankowski AW. *Aspidisca beringiana* sp. nov. and *Simbiodisca* subgen. nov. (Ciliophora: Euplotida), a symbiont of terebellid polychaetes in the Bering Sea. Zoosystematica Rossica. 2009a;18(2): 184-190

[116] Obanda V, Gakuya F, Lekolool I, Chege S, Okita B, Manyibe T. Ciliated intestinal protozoa of Black (*Diceros bicornis michaeli*) and white rhinoceroses (*Ceratotherium simum simum*) in Kenya. African Journal of Ecology. 2007;46: 144-148

[117] Lynn DH, Strüder-Kypke M. Phylogenetic position of *Licnophora*, *Lechriopyla*, and *Schizocaryum*, three unusual ciliates (Phylum Ciliophora) endosymbiotic in echinoderms (Phylum Echinodermata). Journal of Eukaryotic Microbiology. 2002;49(6):460-468

[118] Gürelli G, Lyons ET, Kesbiç FI. Hindgut ciliate composition of throughbred mares in Kentucky, USA, and binary fission in *Polymorphella ampulla*. Zootaxa. 2019;4646(2):369-384

[119] Rataj M, Vďačný P. Living morphology and molecular phylogeny of oligohymenophorean ciliates associated with freshwater turbellarians. Diseases of Aquatic Organisms. 2019;134:147-166

[120] Lynn DH, Wright ADG. Biodiversity and molecular phylogeny of Australian *Clevelandella* species (Class Armophorea, Order Clevelandellida, Family Clevelandellidae), intestinal endosymbiotic ciliates in the wood-feeding roach *Panesthia cribrata* Saussure, 1864. Journal of Eukaryotic Microbiology. 2013;60:335-341

[121] Scheltema RS. Dispersal of the protozoan *Folliculina simplex* Dons (Ciliophora, Heterotricha) throughout the North Atlantic Ocean on the shells of gastropod veliger larvae. Journal of Marine Research. 1973;31(1):11-20

[122] Curds CR. A guide to the species of the genus *Euplotes* (Hypotrichida, Ciliatea), Bulletin of the British Museum (Natural History) Zoology. 1975;28(1): 1-61

[123] Obert T, Vďačný P. Evolutionary origin and host range of *Plagiotoma*

lumbrici (Ciliophora, Hypotrichia), an obligate gut symbiont of lumbricid earthworms. Journal of Eukaryotic Microbiology. 2020;67:176-189

[124] Li M, Ponce-Gordo F, Grim JN, Wang C, Nilsen F. New insights into the molecular phylogeny of *Balantidium* (Ciliophora, Vestibuliferida) based on the analysis of new sequences of species from fish hosts. Parasitology Research. 2014;113:4327-4333

[125] Kittelmann S, Devente SR, Kirk MR, Seedorf H, Dehority BA, Janssen PH. Phylogeny of intestinal ciliates, including *Charonina ventriculi*, and comparison of microscopy and 18S rRNA gene pyrosequencing for rumen ciliate community structure analysis. Applied and Environmental Microbiology. 2015;81:2433-2444

[126] Cameron SL, Wright ADG, O'Donoghue PJ. An expanded phylogeny of the Entodiniomorphida (Ciliophora: Litostomatea). Acta Protozoologica. 2003;42:1-6

[127] Foissner W. Ontogenesis in ciliated protozoa, with emphasis on stomatogenesis. In: Hausmann K, Bradbury PC, editors. Ciliates cells as organisms. Stuttgart: Gustav Fischer Verlag; 1996. p. 95-158

[128] Fahrni JF. Morphologie et ultrastucture de *Spirochona gemmipara* Stein, 1852 (Ciliophora, Chonotrichida). III. Appareil de fixation chez la larve et chez l'adulte. Journal of Protozoology. 1984;31(2):221-232

[129] Bardele CF. Budding and metamorphosis in *Acineta tuberosa*. An electron microscopic study on morphogenesis in Suctoria. Journal of Protozoology. 1970;17(1):51-70

[130] Yi YL, Lu C, Hu XG, Ling F, Wang GX. Antiprotozoal activity of medicinal plants against *Ichthyophthirius multifiliis* in goldfish (*Carassius auratus*). Parasitology Research. 2012;111:1771-1778

[131] Raabe Z. Ordo Thigmotricha (Ciliata-Holotricha). II. Familia Hemispeiridae. Acta Protozoologica. 1970a;7:117-180

[132] Raabe Z. Ordo Thigmotricha (Ciliata-Holotricha). III. Familiae Ancistrocomidae et Sphenopryidae. Acta Protozoologica. 1970b;7:385-463

[133] Raabe Z. Ordo Thigmotricha (Ciliata-Holotricha). IV. Familia Thigmophryidae. Acta Protozoologica. 1971;9:121-170

[134] Raabe Z. Ordo Thigmotricha (Ciliata-Holotricha). V. Familiae Hysterocinetidae et Protoanoplophryidae. Acta Protozoologica. 1972;10:115-184

[135] Lynn DH, Strüder-Kypke, MC. The sanguicolous apostome *Metacollinia luciensis* Jankowski 1980 (Colliniidae, Apostomatia, Ciliophora) is not closely related to other sanguicolous apostomes. Journal of Eukaryotic Microbiology. 2018;66:140-146

[136] Landers SC, Gómez-Gutiérrez J, Peterson WT. *Gymnodinioides pacifica*, n. sp., an exuviotrophic ciliated protozoan (Ciliophora, Apostomatida) from euphausiids of the Northeastern Pacific. European Journal of Protistology. 2006;42:97-106

[137] Lee RF, Walker AN, Landers SC, Walters TL, Powell SA, Frischer ME. Black spot gill syndrome in the Northern shrimp, *Pandalus borealis*, caused by the parasitic ciliate *Synophrya* sp. Journal of Invertebrate Pathology. 2019;161:40-46

[138] Bradbury PC. *Conidophrys pitelkae*, a new species of pilisuctorian from cuticular hairs of *Crangon crangon* (Linneus). Acta Protozoologica. 1975;14(2):161-170

[139] de Padúa SB, Ishikawa MM, Ventura AS, Jeronimo GT, Martins ML, Tavares LER. Brazilian catfish

parasitized by *Epistylis* sp. (Ciliophora, Epistylididae), with description of parasite intensity score. Parasitology Research. 2013;112:443-446

[140] Valladao GMR, Leyva-Pereira N, Viadanna PHO, Gallani SU, Farias THV, Pilarski F. Haematology and histopathology of Nile tilapia parasitised by *Epistylis* sp., an emerging pathogen in South America. Bulletin of the European Association of Fish Pathologists. 2014;35(1):14-20

[141] Saha M, Bandyopadhyay PK, Göcmen B. First record of ectoparasitic ciliates, of genus *Trichodina* (Ciliophora: Trichodinidae) parasiting cultured Oranda Gold fish (*Carassius auratus auratus*) in India. Zootaxa. 2017; 4319(1):128-142

[142] Fariya N, Abidi R, Chauhan UK. Morphological and molecular description of a new species, *Trichodina cirhinii* sp. nov. (Ciliophora: Trichodinidae), infesting native freshwater fish *Cirrhinus mrigala*. Journal of Biological Sciences and Medicine. 2017;3(1):10-17

[143] Akşit D, Falakali-Mutaf B, Göçmen B, Gürelli G. A preliminary observations on *Trichodina* sp. (Ciliophora: Peritricha) on the gills of limpets (*Patella* spp.) in Antalya (Turkey). North-Western Journal of Zoology. 2008;4(2):295-299

[144] van As LL, Basson L, van As JG. Mobiline peritrichs (Ciliophora) collected from the gills of African limpets. Acta Protozoologica. 2017;56:245-254

[145] Wahl M. Marine epibiosis. I. Fouling and antifouling: some basic aspects. Marine Ecology Progress Series. 1989;58:175-189

[146] Wahl M, Mark O. The predominantly facultative nature of epibiosis: experimental and observational evidence. Marine Ecology Progress Series. 1999;187:59-66

[147] Xu Z. The abundance of epizoic ciliate *Epistylis daphniae* related to their host *Moina macrocopa* in an urban stream. Journal of Invertebrate Pathology. 1992;60:197-200

[148] Schuwerack PM, Lewis JW, Jones PW. Pathological and physiological changes in the South African freshwater crab *Potamonautes warreni* Calman induced by microbial gill infections. Journal of Invertebrate Pathology. 2001;77:269-279

[149] Puckett GL, Carman KR. Ciliate epibiont effects on feeding, energy reserves, and sensitivity to hydrocarbon contaminants in an estuarine harpacticoid copepod. Estuaries. 2002;25(3):372-381

[150] Gregori M, Fernandez-Leborans G, Roura Á, González ÁF, Pascual S. Description of a new epibiotic relationship (Suctorian–Copepoda) in NE Atlantic waters: from morphological to phylogenetic analyses. Acta Zoologica. 2016;97(2):165-176

[151] Burris ZP, Dam HG. Deleterious effects of the ciliate epibiont *Zoothamnium* sp. on fitness of the copepod *Acartia tonsa*. Journal of Plankton Research. 2014;36(3):788-799

[152] Fernandez-Leborans G, Gabilondo R. Taxonomy and distribution of the hydrozoan and protozoan epibionts on *Pagurus bernhardus* (Linnaeus 1758) (Crustacea Decapoda) from Scotland. Acta Zoologica. 2006;87(1):33-48

[153] Hanamura Y. Seasonality and infestation pattern of epibiosis in the beach mysid *Archaemysis articulata*. Hydrobiologia. 2000;427:121-127

[154] Utz LRP. Growth of the peritrich epibiont *Zoothamnium intermedium* Precht, 1935 (Ciliophora, Peritrichia) estimated from laboratory experiments. Brazilian Journal of Biology. 2008;68:441-446

[155] Bickel SL, Tang KW, Grossart HP. Ciliate epibionts associated with crustacean zooplankton in German lakes: distribution, motility, and bacterivory. Frontiers in Microbiology. 2012;3(243):1-11

[156] Jones S, Carrasco NK, Perissinotto R, Vosloo A. Association of the epibiont *Epistylis* sp. with a calanoid copepod in the St Lucia Estuary South Africa. Journal of Plankton Research. 2016; 38(6):1404-1411

[157] Utz LRP, Coats DW. Spatial and temporal patterns in the occurrence of peritrich ciliates as epibionts on calanoid copepods in the Chesapeake Bay, USA. Journal of Eukaryotic Microbiology, 2005;52(3):236-244

[158] Souissi A, Souissi S, Hwang JS. The effect of epibiont ciliates on the behavior and mating success of the copepod *Eurytemora affinis*. Journal of Experimental Marine Biology and Ecology. 2013;445:38-43

[159] Mackie RI. Mutualistic fermentative digestion in the gastrointestinal tract: diversity and evolution. Integrative and Comparative Biology. 2002;42:319-326

[160] Schmidt GD, Roberts LS. Foundations of parasitology. 9 ed. New York: The McGraw-Hill Company; 2013. 670p

[161] Vďačný P. Evolutionary associations of endosymbiotic ciliates shed light on the timing of the marsupial-placental split. Molecular Biology and Evolution. 2018;35(7):1757-1769

[162] Corliss JO. The ciliated protozoa; characterization, classification and guide to the literature. 2nd ed. Oxford: Pergamon Press; 1979. 455p

[163] Williams AG, Coleman GS. The rumen protozoa. New York: Springer; 1992. 442p

[164] Kumar S, Pitta DW. Revolution in Rumen Microbiology. In: Puniya AK, Singh R, Kamra, DN, editors. Rumen Microbiology, From Evolution to Revolution. New Delhi: Springer; 2015. p. 357-379

[165] Mizrahi I. Rumen symbioses. In: Rosenberg E, DeLong F, Lory S, Stackebrandt E, Thompson F, editors. The Prokaryotes. Berlin Heidelberg: Springer Science & Business Media; 2013. p. 533-544

[166] Nagaraja TG. Microbiology of the rumen. In: Millen DD, Arrigoni MB, Pacheco RDL, editors. Rumenology. Switzerland: Springer; 2016. p. 39-62

[167] Imai S. Phylogenetic taxonomy of rumen ciliate protozoa based on their morphology and distribution. Journal of Applied Animal Research. 1998;13:17-36

[168] Paul SS, Dey A. Domesticated rare animals (yak, mithun and camel): rumen microbial diversity. In: Puniya AK, Singh R, Kamra DN, editors. Rumen Microbiology: From Evolution to Revolution. New Delhi: Springer; 2015. p. 31-36

[169] Wright ADG, Lynn DH. Maximum ages of ciliate lineages estimated using a small Subunit rRNA molecular clock: crown eukaryotes date back to the Paleoproterozoic. Archiv für Protistenkunde. 1997;148:329-341

[170] Pianka ER. Evolutionary ecology. Sixth edition. San Francisco: Addison Wesley Longman; 2000. 512p

[171] Fenchel T. Ecology of protozoa the biology on free-living phagotrophic protists. New York: Springer; 1987. 197p

[172] Fernandes NM, Schrago CG. A multigene timescale and diversification dynamics of Ciliophora evolution. Molecular Phylogenetics and Evolution. 2019;139:106521

[173] Gentekaki E, Kolisko M, Gong Y, Lynn DH. Phylogenomics solves a long-standing evolutionary puzzle in the ciliate world: the subclass Peritrichia is monophyletic. Molecular Phylogenetics and Evolution. 2017;106:1-5

[174] Vďačný P, Érseková E, Šoltys K, Budiš J, Pecina L, Rurik, I. Co-existence of multiple bacterivorous clevelandellid ciliate species in hindgut of wood-feeding cockroaches in light of their prokaryotic consortium. Scientific Reports. 2018;8(1):1-13

[175] Jackson A. The evolution of parasite genomes and the origins of parasitism. Parasitology. 2015;142:S1-S5

[176] Luong L, Mathot K. Facultative parasites as evolutionary stepping-stones towards parasitic lifestyles. Biology Letters. 2019;15

[177] Poulin R, Randhawa H. Evolution of parasitism along convergent lines: from ecology to genomics. Parasitology. 2013;142:S6-S15

[178] Viney M. How can we understand the genomic basis of nematode parasitism?. Trends in Parasitology. 2017;33(6):444-452

[179] Baker J. The origins of parasitism in the protists. International Journal of Parasitology. 1994;24:1131-1137

[180] Kreier J, Baker JR. Parasitic Protozoa. Winchester, Massachusetts, USA: Allen & Unwin; 1987. 256p

[181] Leung TLF. Fossils of parasites: what can the fossil record tell us about the evolution of parasitism?. Biological Reviews. 2017;92(1):410-430

[182] Rataj M, Vďačný P. Multi-gene phylogeny of *Tetrahymena* refreshed with three new histophagous species invading freshwater planarians. Parasitology Research. 2020;119:1523-1545

[183] Strüder-Kypke M.C, Wright ADG, Jerome CA, Lynn DH. Parallel evolution of histophagy in ciliates of the genus *Tetrahymena*. BMC Evolutionary Biology. 2001;(1)5

[184] Clamp JC, Lynn DH. Investigation the biodiversity of ciliates in the 'Age of Integration'. European Journal of Protistology. 2017;61:314-322

[185] Jamy M, Foster R, Barbera P, Czech L, Kozlov A, Stamatakis A, Bending G, Hilton S, Bass D, Bur F. Long read metabarcoding of the eukaryotic rDNA operon to phylogenetically and taxonomically resolve environmental diversity. Molecular Ecology Resources. 2020;20(2):429-443

[186] Pylro V, Roesch L. The Brazilian microbiome. Current status and perspectives. Switzerland: Springer International Publishing; 2017. 128p

[187] Santoferrara L, Burki F, Filker S, Logares R, Dunthorn M, McManus GB. Perspectives from ten years of protist studies by high-throughput metabarcoding. Journal of Eukaryotic Microbiology. 2020;67:612-622

[188] Mayr E. Systematics and the origin of species from the viewpoint of a zoologist. New York: Columbia University Press; 1942. 334p

[189] Rissler LJ, Apodaca JJ. Adding more ecology into species delimitation: ecological niche models and phylogeography help define cryptic species in the black salamander (*Aneides flavipunctatus*). Systematic Biology. 2007;56:924-942

[190] Knowles LL, Carstens B. Delimiting species without monophyletic gene trees. Systematic Biology. 2007;56:887-895

[191] Pons J, Barraclough TG, Gomez-Zurita J, Cardoso A, Duran DP, Hazell S, Kamoun S, Sumlin, WD,

Vogler AP. Sequence- based species delimitation for the DNA taxonomy of undescribed insects. Systematic Biology. 2006;55:595-609

[192] Yang Z, Rannala B. Bayesian species delimitation using multilocus sequence data. PNAS. 2010;107:9264-9269

[193] Zhang J, Kapli P, Pavlidis P, Stamatakis A. A general species delimitation method with applications to phylogenetic placements. Bioinformatics. 2013;29:2869-2876

[194] Zhao Y, Yi Z, Gentekaki E, Zhan A, Al-Farraj SA, Song W. Utility of combining morphological characters, nuclear and mitochondrial genes: an attempt to resolve the conflicts of species identification for ciliated protists. Molecular Phylogenetics and Evolution. 2016;94:718-729

[195] Shazib SUA, Vd'acˇny'P, Kim JH, Jang SW, Shin MK. Molecular phylogeny and species delimitation within the ciliate genus *Spirostomum* (Ciliophora, Postciliodesmatophora, Heterotrichea), using the internal transcribed spacer region. Molecular Phylogenetics and Evolution. 2016;102:128-144

[196] Mehlhorn H. Animal Parasites. Diagnosis, treatment, prevention. Switzerland; Springer International Publishing; 2016. 730p

[197] Ohtsuka S, Suzaki T, Horiguchi T, Suzuki N, Not F. Marine Protists. Diversity and dynamics. New York: Springer; 2015. 637p

Chapter 8

Changes in the Fatty Acids Profile of the Zooplankton Community Reveals the Quality of Four Reservoirs in the Hydroelectric Power Plants Located in the Iguaçu River, Paraná, Brazil

Fabiano Ramiro Serpe, Júlio César Rodrigues de Azevedo, Fernando J.M. Gonçalves, João C. Marques and Ana M.M. Gonçalves

Abstract

Fatty acids are molecules with important physiological functions, proved to be good bioindicators of the presence of natural and chemical stressors and so used as early warning signals. Indeed, biochemical analyzes, such as fatty acids, are an important tool in water body management and water quality analysis, allowing detecting molecular changes in aquatic communities, related to the trophic status of the systems, before they are perceived in the environment. In this work was investigated the fatty acid composition on zooplankton community collected in four reservoirs of hydroelectric plants on the Iguaçu River, Brazil, and assessed the species distribution to assess and compare the water quality in these reservoirs. Results showed the trophic state index presented a wide variation among samples, ranging from oligotrophic (Salto Caxias) to hypereutrophic (Foz do Areia). The most abundant fatty acid was docosahexaenoic acid (DHA, C22:6n3) an essential fatty acid with health benefits, playing a pivotal role in biological functions. This study highlights the sensitiveness of the zooplankton community to environmental conditions and underlines the role of fatty acids as good bioindicators, being good endpoints to use in ecological studies. This supports the zooplankton contribution as a biological quality element in the assessment of reservoir quality elements.

Keywords: fatty acids, hydroelectric reservoirs, Iguaçu river, reservoir dynamics, zooplanktonic diversity

1. Introduction

The development of urban centers leads large cities in many regions (for example, as in South America and Asia) to discharges of industrial and domestic wastewater treated inadequately or depleted directly to the environment. The release of sewage from residential and industrial areas and the overuse of fertilizers and pesticides in aquatic environments, cause deoxygenation, increases the levels of toxic compounds and spread infectious diseases, with the degradation of water quality and significant negative impacts on health and mortality [1]. In addition, overexploitation of groundwater can damage wetlands, cause soil subsidence, and induce saltwater intrusion into coastal aquifers. In other regions, over-exploitation of surface water disrupts flow regimes, affecting aquatic ecosystems and the quantity and quality of water supply. Thus, extensive hydrological information and monitoring ecosystem plans are crucial for the development and protection of water resources. In studies of water quality assessment, a range of parameters are determined and assessed, which can be divided into three groups: (1) analysis of biological parameters (2) determination of physical, chemical, trophic, ecological and saprobicity indices (3) ecological aspects of community's biological processes [2, 3]. Several works were carried out based on the response of specific biological species to determine water quality [3–6]. With the implementation of the European Water Framework Directive (WFD), the ecological status of surface water is classified to standardize procedures based on the evaluation of a series of biological quality elements (BQEs). Different BQEs can act as pressure-respondents with complementary roles as proxies of structural and functional ecology [7]. In lakes, phytoplankton is the "fast responder" to eutrophication, while other BQEs are more sensitive to other pressures like hydromorphological or chemical ones [8–10]. However, zooplankton has not yet been included as a BQE, despite being considered as a key component of aquatic food chains, but the reason for this omission remains unclear.

According to Jeppesen et al. [11], the value of zooplankton as an indicator of ecological conditions stems from its position in the food chain, controlled by top-down regulators (fish) and bottom-up factors (phytoplankton), thus providing information on the relative importance of both main regulatory processes, as well as the impact of zooplankton on water quality. These authors concluded that the focus primarily on ecosystem structure and less on WFD should be reconsidered, and it should be demonstrated that zooplankton is a key element in understanding the function of lake ecosystems and perhaps also in large rivers and transitional waters [11].

The use of zooplankton for the environmental characterization of aquatic environments is potentially advantageous because of its key position in the food chain, and for wide geographic distributions [12, 13]. These organisms respond rapidly to acute and chronic stress factors, showing a high sensitiveness to chemical and environmental stressors, making them favorable candidates as indicators of ecosystem quality [2, 3].

The concentration of nutrients varies among the various aquatic systems influencing the chemical composition of the aquatic organisms. Some organisms are considered good bioindicators and can be used in studies of water quality, since their occurrence is related to the degree of pollution of the sampling site [3, 5]. Thus, observing the variations in the biochemical composition of the zooplankton organisms, can be correlated it with the eutrophication status of the system, and infer about the quality of the ecosystem. Indeed, lipids are very sensitive to environmental and chemical stressors [12, 14, 15]. In the last decades, the interest in the fatty composition of aquatic organisms has increased. The knowledge in biochemical composition of the main zooplankton groups has become important to understand the metabolism, physiological functions and nutritive value due to its relevance for the energy transfer in aquatic systems and secondary production.

Still, studies on the biochemical composition of zooplankton, more precisely on fatty acids, related to the trophic condition of the environment are scarce in the literature.

In Brazil, the reservoirs of the five hydroelectric plants located on the Iguaçu River—Foz do Areia, Salto Segredo, Salto Santiago, Salto Osório and Salto Caxias - are characterized by forming a cascade system. The mainstream of Iguaçu River and some of its tributaries are currently polluted and receive high man-induced loads of nutrients, substantially originating from domestic sewage. This study aims to determine the fatty acids composition of the zooplankton community collected from four reservoirs along the Iguaçu River, and to relate them to the ecological status of the aquatic system to assess its quality.

2. Material and methods

2.1 Study area

The Iguaçu River originates from the junction of the Iraí and Atuba rivers, in the metropolitan region of Curitiba, Paraná State. Its formations, originating at altitudes above 1000 m, constitute the Iguaçu River at an elevation of 908 m, from where it travels 1060 km, in the east-west direction, receiving water from various tributaries until reaching an altitude of 78 m and flowing into the Paraná River, near the city of Foz do Iguaçu [16]. From the Paraná rivers, it has the largest hydrographic basin, covering an area of approximately 72,000 km^2, of which 79.00% belong to the state of Paraná, 19.00% to the state of Santa Catarina and 2.0% to Argentina [17]. The Iguaçu River is the main river in the State of Paraná, runs from east to west, having its source located near the municipality of Curitiba and its mouth in the city of Foz do Iguaçu.

Due to the favorable conditions of uneven terrain, several hydrographic basins (among them that of the Iguaçu River) were used for the construction of reservoirs in sequence. The series of dams built in the same hydrographic basin forms what is known as a cascade of reservoirs [18], a condition that changed the physiography in many hydrographic basins in the country.

There are five large reservoirs for power generation along the Iguaçu River, located in the southern region of Brazil, in the Paraná state, all with more than 80 km^2 of surface area: Foz do Areia, Salto Segredo, Salto Santiago, Salto Osório and Salto Caxias (**Figure 1**). Together they have a surface area of 753.98 km^2 and an installed power generation capacity of 6644 megawatts, contributing to 6.54% of the national production. In general, they are dendritic and deep reservoirs, Foz do Areia has 180 m deep [16]. These reservoirs, built in a cascade system are usually operated as single units, so from the physical, chemical and biological point of view, each can behave as a unit with unique characteristics [19].

The Foz do Areia reservoir is the first of the large reservoirs of the Iguaçu River, it was formed in 1980 by a 160 m high and 820 m long dam, flooding an area of 139 km^2 on the border between the municipalities of Pinhão and Bituruna. The reservoir has its margins protected by natural vegetation and regions with secondary forests, mainly due to the relief of the region. The banks of Foz do Areia reservoir are constituted of natural vegetation and agricultural lands [20].

The Salto Segredo reservoir is located downstream of the Foz do Areia reservoir and upstream of the Salto Santiago reservoir, in the municipalities of Reserva do Iguaçu and Mangueirinha, was formed in 1992, with a flooded area of 82.5 km^2. It is

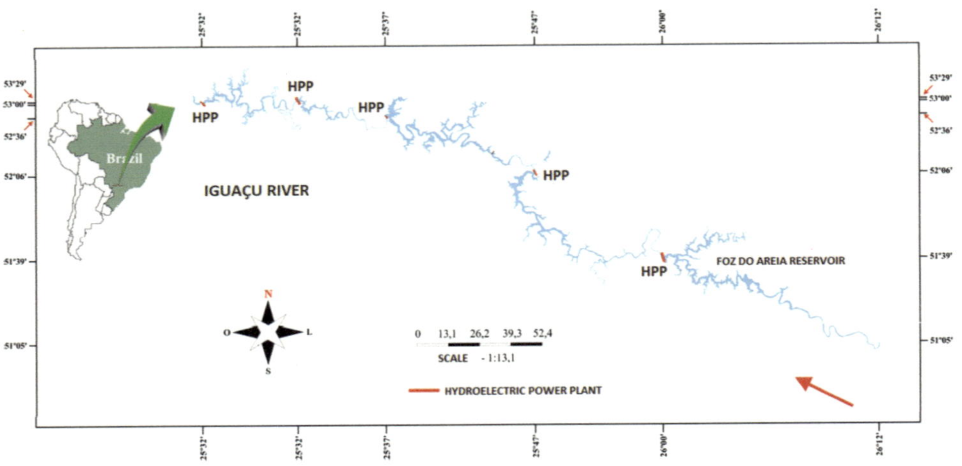

Figure 1.
Reservoirs along the Iguaçu River (modified from da Silva [19]).

a little dendritic reservoir with an average depth of 36.6 m (in some places it can reach up to 100 m) and a water residence time of 47 days. It receives numerous tributaries, both on its right bank (Forest, São Pedro, Verde and Touros) and on its left bank (Patos, Iratim, Butiá and Covó). In Salto Segredo reservoir there is an Environmental Protection Area (APA). The area that is not covered by the protection area is of agricultural use [20].

The closure of the Salto Santiago dam took place in 1979, flooding an area of 208 km^2, on the border between the municipalities of Rio Bonito do Iguaçu and Saudade do Iguaçu. The main dam is 80 m high and 1400 m long. This reservoir covers an area of 208 km^2 and covers part of the territory of eight municipalities, five on the right bank—Rio Bonito do Iguaçu, Porto Barreiro, Virmond, Candói and Foz do Jordão (central west region)—and three on the left bank—Saudade do Iguaçu, Chopinzinho and Mangueirinha (southwest region).

The Salto Santiago Lake differs from the others for its size, the great distance between its banks, the large meanders, the large number of inlets on the edges and, above all, the predominance of immense vertical walls on its banks [21].

Salto Caxias is the last of the large reservoirs on the Iguaçu River, it was closed in 1998, flooding an area of 124 km^2 on the border of the municipalities of Capitão Leônidas Marques and Nova Prata do Iguaçu. The relief of the area surrounding the reservoir is less accentuated than that of the Segredo region, with more intense agricultural occupation, being dominated by pastures with small areas of secondary forests [20]. In Salto Santiago and Salto Caxias reservoirs the banks are constituted mainly of agricultural lands. All these reservoirs are artificial, built by man, to generate energy.

It was not possible to carry out sampling in the Salto Osório reservoir due to logistical issues, with this reservoir not included in the study. According to the Paraná Reservoirs Water Quality Report (2005–2008), the water quality of the Foz do Areia reservoir is the worst among the plants on the Iguaçu River. This reservoir was considered moderately degraded. The other reservoirs showed better qualities, with Salto Segredo considered moderately degraded, and Salto Santiago, Salto Osório and Salto Caxias classified as little degraded. From the report, it is possible to observe a gradual improvement in reservoirs further away from the metropolitan region of Curitiba. When passing through this region, the Iguaçu River receives a large amount of polluting substances of domestic, industrial and diffused origin, most of which are untreated, as this is in a region with little access to basic sanitation, as well as the presence of industries. Thus, the pollution load from the metropolitan region of Curitiba (RMC) significantly influences the water quality of the Iguaçu River [22].

2.2 Collection of biotic and abiotic data

Four campaigns were carried out in the four reservoirs: the first in July 2012, the second in November 2012, the third in February 2013 and the last in October 2013. These campaigns were named according to the seasons, the first "W" (Winter), the second "Sp" (Spring 2012), the third "S" (Summer) and the latter "Sp2" (Spring 2013). For logistical reasons, it was not possible to perform sampling in autumn.

To study the spatial distribution and variation of the zooplankton community in the reservoirs of hydroelectric plants, nine collecting stations were chosen, three in the lake region, three in the intermediate region of the lake and three in the river region. In each region, sampling was carried out in three points: one in the center of the lake and at two points in the lake shores.

The collections were carried out during the period between 8:00 AM and 2:00 PM in all campaigns carried out, during high tide, maintaining the same conditions in all sampling collections. At each sampling site, two samples were collected, one of which was placed in a specific flask and preserved in 4% formaldehyde saturated with sucrose [23], to prevent distortion of the shells and loss of eggs in Cladocera. These samples were counted and analyzed using an optical microscope with a Sedgwick-Rafter counting chamber. The other sample was preserved on ice and transported to the laboratory, whereas much water as possible was removed, frozen and subsequently lyophilized.

For the qualitative and quantitative analysis of the zooplankton community were performed vertical hauls on the water surface to a 0.3 m depth, with zooplankton nets with 45 μm mesh opening and 0.3 m of mouth diameter. The tows were performed on the boat with an electric motor at a two-nodes speed, for 5 min. The total filtered water is calculated using the cylinder volume formula, and the final volume of each sampling was about 13 m^3.

In all collections, the following environmental variables were measured: water temperature (°C), dissolved oxygen concentration (mg L^{-1}) and pH, using a multiparameter probe. Water samples (1000 mL) were also collected at all collection points for further quantification of dissolved organic carbon (DOC), phosphorus forms (total phosphorus (P-total) and phosphate ($P\text{-}PO_4^{3-}$)), the forms of nitrogen (nitrite (NO_2^-), ammonia nitrogen ($N\text{-}NH_3$), nitrate (NO_3^-), organic nitrogen (N-org), total nitrogen (N-total)) were determined by the methods described in APHA [24].

2.3 Calculation of trophic state index (TSI)

To calculate the trophic state index of the reservoir, the trophic state index for chlorophyll (TSI_{CL}) and the trophic state index for phosphorus (TSI_P) were initially determined.

The trophic state indices (TSI_{CL} and TSI_P) were calculated according to Lamparelli [25]. In reservoirs, the calculation of the TSI from phosphorus values is performed by the equation (1),

$$TSI = 10.\left(6 - \left(\frac{1,77 - 0,42.(\ln TP)}{\ln 2}\right)\right) \tag{1}$$

the concentration of total phosphorus (TP) is expressed in μg L^{-1}.

The calculation of the TSI from the chlorophyll values is performed by the equation (2),

$$TSI = 10.\left(6 - \left(\frac{0,92 - 0,34.(\ln Cl)}{\ln 2}\right)\right) \quad (2)$$

the concentration of chlorophyll (Cl) is expressed in μg L^{-1}. The chlorophyll concentration was quantified by the spectrophotometric method described in APHA [24].

The TSI is the result of the arithmetic mean between the TSI_{CL} and the TSI_P.

2.4 Determination of fatty acid profile

Zooplankton organisms collected in the field were previously frozen, lyophilized, placed in Eppendorf's and later maintained with silica gel in the freezer, to avoid lipid photooxidation [26] and subsequently frozen at −80°C. For each sampling site, three replicates were prepared and weighed.

The extraction of total lipids from the zooplankton community and the methylation of fatty acid methyl esters (FAMEs) for fatty acid analysis was performed as described by Gonçalves et al. [12]. Samples were incubated with methanol for the extraction of lipids. The fatty acid Methylnonadecanoate (C19:0) was added as an internal standard for quantification. The samples were centrifuged and dried under a vacuum. The FAMEs obtained were analyzed by a Trace 1300 ThermoScientific GC. The GC used a FAME biodiesel column (60 m × 0.250 mm × 0.20 μm). The column temperature was programmed to increase from 140 to 240°C, the analysis time was 45 min per sample, injecting 1.0 μl of the sample, and the carrier gas was Helium (20 cm/s, 175°C).

The FAMEs were identified by comparison with the retention times. The quantification of the individual FAMEs was performed by external standards, and the function of the quantification of each FAME was obtained by linear regression applied to the areas of the chromatographic peaks and corresponded with the known concentrations of the standards [12].

In fatty acid data analyses, an average of the sampling points to each region was calculated, because in some samples of some collection points it was not possible to extract the fatty acids, preventing the comparison with the results of other samplings. Region 1 (R1) corresponded to the region farthest from the dam, region 2 (R2), corresponded to the central region of the lake, and region 3 (R3) corresponded to the region near the dam of the reservoir.

2.5 Statistical analysis

Multivariate statistical analysis was carried out to examine the variation in fatty acid profiles, spatial and temporally, through multidimensional scaling (n-MDS) plots. Cluster analysis using the Bray-Curtis dissimilarity factor, using the group average was performed to assess the degree of similarity between the fatty acid samples. For these multivariate analyzes, the program PRIMER-E 6 was used. Data was not normalized, as the values are very similar, the Kruskal adjustment scheme was used, adopting 25 restarts and minimum stress of 0.01.

To assess the temporal and spatial changes of the physical, chemical and biological variables, data were processed from a matrix using Principal Component Analysis (PCA), which was based on eigenvalues greater than 1.0, which explained 70% of the total variability. To obtain greater reliability in the data analysis, greater importance was given to the correlations between factors and variables greater than or equal to 0.7.

3. Results

3.1 Abundance of zooplankton community

In the Foz do Areia reservoir, 35 species of rotifers, 11 species of cladocerans and 3 species of copepods were found throughout the study period (**Table 1**). Although, *Polyarthra dolichoptera* was the most abundant rotifer during the sampling in July 2012 (W) and spring 2013 (Sp2), the abundance averages were quite different, from 5978.58 to 537.48 ind m^{-3}, respectively. In addition, species' abundance was also differentiated, with 20 species in July 2012 (W) and 10 species in October 2013 (Sp2) collections. In spring 2012 (Sp—November 2012) and summer 2013 (S—February 2013) different abundances of rotifer species were registered, with 16 and 23 species respectively, with *Synchaeta jollyae* being the most abundant species in spring 2012 (average density of 1427.50 ind m^{-3}), and *Keratella cochlearis* the most abundant rotifer species in spring 2012 (average density: 1628.33 ind m^{-3}) (**Table 2**). The presence of several zooplankton species in the Foz do Areia reservoir may be related to the higher effluent load, as it is the first dam after the metropolitan region.

The abundance of cladocerans species in winter (W—July 2012) and in summer (S—February 2013) was the same, with *Ceriodaphnia cornuta* being the most abundant cladoceran (256.83 ind m^{-3} in winter and 106, 33 ind m^{-3} in the summer). In spring of 2012 (Sp—November 2012) and 2013 (Sp—October 2013) there was an equal abundance of species, with *Moina minuta* being the most abundant in spring 2012 (average density: 464.17 ind m^{-3}), and *Bosmina longirostris* the most abundant cladoceran species in spring 2013 (average density: 171.40 ind m^{-3}). Among copepods, naupliar stages were the most abundant (**Table 2**).

In the Salto Segredo reservoir 28 species of rotifers, 10 species of cladocerans and two species of copepods were found throughout the study period. The most abundant rotifer was *P. dolichoptera* in three of the four samples, November 2012 (Sp), February 2013 (S) and October 2013 (Sp2) with density values of 19.22, 1011.67 and 748.10 ind m^{-3}, respectively. Although, *P. dolichoptera* was the most abundant organism during these periods, the species abundance was different, with 9 (Sp), 13 (S) and 11 (Sp2) species (**Table 2**).

In July 2012 sampling (W) 17 species of rotifers were found, in which the most abundant was *K. cochlearis* with an average of 640.00 ind m^{-3}. Regarding cladocerans, *B. longirostris* was the most abundant species (156 ind m^{-3}) in July 2012 (W), *C. cornuta* was the most abundant cladoceran (120.22 ind.m^{-3}) in November 2012 (Sp), *Diaphanosoma spinulosum* was the most abundant cladoceran (210,44 ind m^{-3}) in February 2013 (S) and *Bosminopsis deitersi* the most abundant species in the October 2013 sampling (Sp2). In spring of 2012 (Sp—November 2012) nine cladoceran species were found, with *B. longirostris* being the most abundant, with an average of 156.00 ind m^{-3}, and in the summer of 2013 (S—February of 2013) six species of cladocerans were found, and the most abundant was *D. spinulosum* with a mean density of 210.44 ind m^{-3} (**Table 2**).

Regarding copepods, in all samples only two species of copepods were found, with the exception of the October 2013 sampling, where only one species was found. In all samples, the naupliar stages were the most abundant organisms.

In the Salto Santiago reservoir, in July 2012 sampling (W) 19 species of rotifers were identified, in which the most abundant was *Kellicottia longispina* with an average of 120.00 ind m^{-3}; in November 2012 sampling (Sp), were identified 13 species of

	Rotifera		Reservoirs			
Order		**Code**	**FA**	**SG**	**ST**	**CX**
Bdelloida (*)	*Bdelloida* (*)	Bdel	X	X	X	
Family	**Species/Genus**					
Asplanchnidae	*Asplanchna* sp.	Aspl	X	X	X	X
Brachionidae	*Anuraeopsis fissa*	Ap fiss	X			
	Brachionus caudatus	Brach cd	X			
	Brachionus dolabratus	Brach dl	X	X	X	X
	Brachionus falcatus	Brach fc	X		X	X
	Brachionus sp.	Brach	X			
	Brachionus urceolaris	Brach urc	X			
	Kellicottia longispina	K long	X	X	X	X
	Keratella americana	K amer	X	X		
	Keratella cochlearis	K coch	X	X	X	X
	Keratella sp.	K sp	X			
	Keratella tropica	K trop	X	X		X
	Keratella valga	K val	X			
	Notholca sp.	Noth	X	X		
	Platyas quadricornis	P quadr		X	X	
Collothecidae	*Collotheca ornata*	Cornt	X	X	X	X
	Collotheca sp.	Collot.	X	X	X	X
	Colurella sp.	Colur		X		
Conochilidae	*Conochiloides* sp.	Cchldes	X	X		
	Conochilus sp.	Cchilus	X	X	X	
	Conochilus coenobasis	Cchilus cb	X		X	X
Epiphanidae	*Epiphanes macrourus*	E macr	X		X	
Euchlanidae	*Euchlanis dilatata*	E dilat	X	X	X	X
Filinidae	*Filinia longiseta*	F long	X		X	
	Filinia opoliensis	F opo	X			
	Filinia terminalis	F term	X	X		
Flosculariidae	*Ptygura libera*	Pty lib	X		X	X
Gastropodidae	*Ascomorpha ovalis*	A ov	X	X	X	X
	Ascomorpha saltans	A salt			X	
Hexarthridae	*Hexarthra mira*	Hex m	X	X	X	X
Lecanidae	*Lecane bulla*	L bul			X	
	Lecane luna	L luna		X		
	Lecane sp.	L sp.	X	X	X	X
Philodinidae	*Philodina* sp.	Phil		X		
Synchaetidae	*Polyarthra dolichoptera*	P doli	X	X	X	X
	Synchaeta jollyae	Synch jo	X	X	X	X
	Synchaeta sp.	Synch sp.	X	X	X	X

	Rotifera		Reservoirs			
Testudinellidae	*Pompholyx sulcata*	Pom sul		X		X
Trichocercidae	*Trichocerca bicristata*	T bicr		X	X	
	Trichocerca bidens	T bid		X	X	X
	Trichocerca cylindrica	T cylin		X	X	X
	Trichocerca rattus	T rat			X	
	Cladocera					
Family	Species/Genus	Code				
Bosminidae	*Bosmina hagmanii*	Bn hag	X	X	X	X
	Bosmina longirostris	Bn long	X	X	X	X
	Bosminopsis deitersi	Bs deit	X	X	X	X
Chydoridae	*Alona* sp.	Al sp.	X		X	X
	Chydorus sp.	Chyd			X	
	Pseudochydorus globosus	Pschy glob		X	X	
Daphniidae	*Ceriodaphnia cornuta*	Cd corn	X	X	X	X
	Ceriodaphnia silvestrii	Cd silv	X	X	X	X
	Daphnia gessneri	Dp gess	X	X	X	X
	Daphnia laevis	Dp lvis	X	X	X	X
	Daphnia parvula	Dp par	X		X	
Moinidae	*Moina minuta*	Mn min	X	X	X	X
Sididae	*Diaphanosoma spinulosum*	Dph spin	X	X	X	X
	Copepoda					
Order	Species	Code				
Calanoida	Calanoida sp.	Calan sp	X			
	Notodiaptomus spinuliferus ♀	Nt spinF	X	X	X	X
	Notodiaptomus spinuliferus ♂	Nt spinM	X	X	X	X
	Calanoida copepodite	Cp Calan	X	X	X	X
Cyclopoida	Cyclopoida copepodite	Cp Cyclo	X	X	X	X
	Tropocyclops prasinus ♀	Tp prasF	X	X	X	X
	Tropocyclops prasinus ♂	Tp prasM	X	X	X	X
	Nauplius	Naup	X	X	X	X

Table 1.
List of zooplankton taxa found in the Iguaçu River reservoirs during the study period. (*) order. FA = Foz do Areia reservoir, SG = Salto Segredo reservoir, ST = Salto Santiago reservoir, CX = Salto Caxias reservoir.

rotifers, and the most abundant species was *Collotheca* sp. With an average density of 647.78 ind m^{-3}. In samples collected in February 2013 (S), 12 species of rotifers were identified. The most abundant among the rotifers species was *Colurella* sp. With an average of 1164.67 ind m^{-3}. In samples collected in Salto Santiago in October 2013 (Sp2), six species of rotifers were identified, and the most abundant was *Asplanchna* sp. With an average density of 4536.74 ind m^{-3} (**Table 2**).

During July 2012 sampling (W) nine cladoceran species were identified. The most abundant species was *Ceriodaphnia silvestrii* with an average of 140.00 ind m^{-3}, while in the November 2012 sampling (Sp) eight cladoceran species were

	W	Sp	S	Sp2
FOZ DO AREIA reservoir				
Polyarthra dolichoptera	**5978.58**	275.83	332.33	**537.48**
Synchaeta jollyae	0.00	**1427.50**	900.83	0.00
Keratella cochlearis	1083.50	340.00	**1628.33**	66.70
Ceriodaphnia cornuta	**256.83**	170.83	**106.33**	64.36
Diaphanosoma spinulosum	97.50	**464.17**	21.17	5.17
Bosmina longirostris	118.25	149.17	57.50	**171.40**
Nauplii	**8635.83**	**2338.33**	**3130.25**	**2779.07**
SALTO SEGREDO Reservoir				
Keratella cochlearis	**640.00**	9.33	192.33	66.70
Polyarthra dolichoptera	145.89	**19.22**	**1011.67**	**537.48**
Bosmina longirostris	**156.00**	22.00	60.67	107.32
Ceriodaphnia cornuta	36.67	**120.22**	23.89	3.77
Diaphanosoma spinulosum	21.56	3.22	**210.44**	66.53
Bosminopsis deitersi	115.56	1.11	0.00	**118.23**
Nauplii	**1582.22**	**138.11**	**2109.11**	**4356.47**
SALTO SANTIAGO Reservoir				
Kellicottia longispina	**120.00**	2.22	0.00	0.00
Polyarthra dolichoptera	112.22	**355.89**	128.78	217.90
Colurella sp.	0.00	0.00	**1164.67**	0.00
Asplanchna sp.	0.00	0.00	0.00	**4536.74**
Ceriodaphnia silvestrii	**140.00**	156.89	**78.44**	65.40
Ceriodaphnia cornuta	22.22	**804.78**	18.44	0.00
Bosmina longirostris	77.88	96.00	17.44	**171.44**
Nauplii	**254.44**	**948.67**	**703.33**	**2506.94**
SALTO CAXIAS Reservoir				
Polyarthra dolichoptera	**837.78**	406.44	84.86	116.06
Collotheca sp.	250.00	**808.00**	74.36	135.71
Synchaeta sp.	0.00	20.44	**1798.20**	0.00
Asplanchna sp.	0.00	0.00	1.24	**3018.71**
Ceriodaphnia silvestrii	**224.44**	**418.67**	18.06	**100.02**
Diaphanosoma spinulosum	48.56	74.89	**101.57**	18.34
Nauplii	**995.56**	**4473.56**	326.56	**2436.87**
Calanoid Copepodites	524.44	1646.89	**497.90**	91.02

Values in bold in the table indicate the most abundant species.

Table 2.
List of the most abundant taxa *in the Iguaçu River reservoirs.*

found and the most abundant was *C. cornuta*, with an average of 804.78 ind m^{-3}. Seven species of cladocerans were identified in the February 2013 sampling (S) and the most abundant species was *Ceriodaphnia silvestrii*, with a mean density of 78.44 ind m^{-3}. In October 2013 sampling (Sp2), five cladoceran species were identified, and the most abundant cladoceran was *B. Longirostris*, with a mean density of 171.44 ind m^{-3}. Regarding copepods, only two species were found in all samples, with naupliar stages being abundant throughout the study period (**Table 2**).

In the Salto Caxias reservoir, 20 species of rotifers, ten species of cladocerans and two species of copepods were found.

In the July 2012 sampling (W) eight species of rotifers were identified, in which *P. dolichoptera* was the most abundant species with an average density of 837.78 ind m^{-3}. Both in November 2012 (Sp) and February 2013 (S) samplings, the abundance of species was the same (14 species), being *Collotheca* sp. The most abundant in November 2012 (Sp), with an average density of 808.00 ind m^{-3}. In February 2013 (S) *Synchaeta* sp., was the most abundant with an average density of 1798.20 ind m^{-3} (**Table 2**). In October 2013 (Sp2), five species of rotifers were identified with *Asplanchna* sp. The most abundant, with an average density of 3018.71 ind m^{-3}. Except in the sampling of July 2012 (W), where eight cladoceran species were identified, the most abundant species was *Ceriodaphnia silvestrii*, with an average density of 224.44 ind m^{-3}. The other three samples had the same abundance of cladoceran species (eight species).

In November 2012 sampling (Sp), the most abundant species was *Ceriodaphnia silvestrii*, with an average density of 418.67 ind m^{-3}. In February 2013 (S), the most abundant was *D. spinulosum*, with an average density of 101.57 ind m^{-3}, and in the sampling of October 2013 (Sp2), the most abundant species was *Ceriodaphnia silvestrii*, with an average density of 100.02 ind m^{-3}. In all samplings carried out, the naupliar stages were the most abundant among the copepods, with the exception of the February 2013 sampling (S) where the calanoid copepodites were the most abundant, with an average density of 497.90 ind m^{-3} (**Table 2**).

3.2 Trophic state index (TSI)

The TSI showed a wide variation from sampling to sampling. In July 2012 (W) the environment was characterized as oligotrophic, while in the following sampling there was a sharp drop in water quality, and the environment was characterized as hypereutrophic and the following two as eutrophic. The eutrophication process that occurred in November 2012 (Sp), may have been possibly caused by the great drought that occurred at the time of the study, but just before this great drought, there was a large precipitation phase, which may have caused the entry of matter causing the concentration of nutrients that increased temporarily. When the calculated averages were observed in each sampling site, it was seen that the variation was low, with some collected areas with different trophic classifications (**Figure 2**).

3.3 Fatty acids composition of zooplankton community in the sampling sites

In Foz do Areia reservoir, 19 FA were determined during the entire study period, of which nine were saturated fatty acids (SFA), four monounsaturated fatty acids (MUFA), three polyunsaturated fatty acids (PUFA) and three highly unsaturated fatty acids (HUFA). The period in which a greater abundance of fatty acids was observed was in July 2012 (W), in which thirteen fatty acids were identified, mostly SFA, however, in the November 2012 (Sp2) sampling, only seven fatty acids were identified (**Table 3**).

The most abundant fatty acid was docosahexaenoic acid (DHA), with the exception of February 2013 sampling (S), where palmitic acid (C16: 0) was the only one that contributed to the total densities of fatty acids (100%). In the first sampling (W) there was an increase in the number of fatty acids from region 1 (R1) at the region 2 (R2), which covers the points furthest from the dam, and the

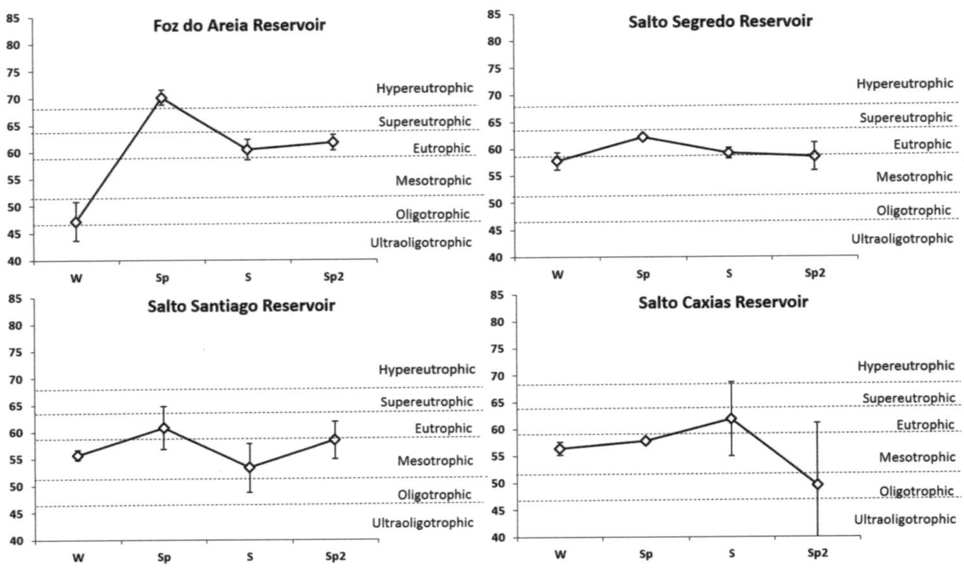

Figure 2.
Average of trophic state indexes for the Iguaçu River reservoirs. The bars indicate the standard deviation.

intermediate points of the reservoir respectively. At the other samples, region 3 (R3), which covers the points closest to the dam, was registered the highest fatty acid densities. Sampling in October 2013 registered the highest number of fatty acids. In region 2, sampling in November 2012 (Sp), DHA represented more than 91% of all identified fatty acids (**Figure 3**).

In Salto Segredo reservoir, 20 fatty acids were found, of which eight are saturated fatty acids (SFA), five monounsaturated fatty acids (MUFA), four polyunsaturated fatty acids (PUFA) and three highly unsaturated fatty acids (HUFA) (**Table 3**). In the two spring samplings (Sp and Sp2) more fatty acids were found, 15 and 16 respectively, of which eight are SFA found in November 2012 sampling (**Table 3**).

The most abundant fatty acid in the Salto Segredo reservoir was docosahexaenoic acid (DHA), representing 96% of the total concentration of fatty acids in the first sampling. The high concentration of this fatty acid appears to be due to the large presence of phytoplankton in the environment, which can be observed by the relatively high concentration of chlorophyll-*a* in the reservoir (on average 5.91 $\mu g\ L^{-1}$). However, in the February 2013 sampling (S) almost no fatty acids were found in the reservoir, and only palmitoleic acid (C16:1) was identified. EPA and C16:0 were also quite representatives (around 20% in both) (**Figure 3**). In a more general context, high C16:1 value indicates a high presence of diatoms [27].

In Salto Santiago reservoir, 18 fatty acids were identified, being six saturated fatty acids (SFA), six monounsaturated fatty acids (MUFA), three polyunsaturated fatty acids (PUFA) and three highly unsaturated fatty acids (HUFA) (**Table 3**). In October 2012 sampling (Sp2), 14 fatty acids were identified, being five SFA and five MUFA; in the February 2013 sampling (S), only three fatty acids were identified, two being SFA and one HUFA.

In Salto Santiago reservoir, DHA was the most abundant fatty acid in the three regions of the samplings of July 2012 (W), February 2013 (S) and October 2013 (Sp2). In the November 2012 sampling (Sp) the fatty acid with the highest density was C16:0, with more than 60% of the total densities (**Figure 3**). Looking at the results, C16:0 seems to be related to poorer water quality, which suggests the algae that is responsible for the production of this fatty acids is probably more abundant

		FOZ DO AREIA				SALTO SEGREDO				SALTO SANTIAGO				SALTO CAXIAS			
		ReservoirS															
F. A	**C:D**	**W**	**Sp**	**S**	**Sp2**	**W**	**Sp**	**S**	**Sp2**	**W**	**Sp**	**S**	**Sp2**	**W**	**Sp**	**S**	**Sp2**
SFA	C8:0									9.24E-05							
	C12:0							8.46E-06									
	C13:0	1.69E-05															
	C14:0	1.75E-03	1.44E-03	9.79E-04	3.08E-03		9.73E-04		7.50E-03	6.09E-05	3.00E-05		5.51E-03	5.84E-03	7.63E-05	2.51E-04	1.57E-02
	C15:0	1.09E-04															
	C16:0	1.51E-02	7.76E-03	8.21E-03	2.16E-02	6.20E-04	8.63E-03		3.89E-02	4.66E-03	4.21E-03	1.99E-04	2.29E-02	4.30E-02	4.45E-03	7.77E-03	7.72E-02
	C17:0				1.94E-04		1.82E-04										
	C18:0	2.91E-03	7.19E-04	7.25E-04	2.46E-03	1.42E-04	1.62E-03		1.92E-03	3.63E-04		2.18E-06	6.66E-04	1.39E-02	5.37E-04	1.30E-03	7.20E-03
	C20:0	7.08E-05		1.75E-03	1.28E-03		6.60E-04						1.37E-04			8.37E-04	
	C21:0			1.45E-03	9.26E-03		4.76E-04		1.75E-02				8.82E-03				1.20E-02
	C22:0	8.22E-04			2.98E-03		2.54E-04		3.39E-03								
	Σ SFA	**2.08E-02**	**9.92E-03**	**1.31E-02**	**4.09E-02**	**7.62E-04**	**1.28E-02**		**6.92E-02**	**5.18E-03**	**4.24E-03**	**2.01E-04**	**3.80E-02**	**6.27E-02**	**5.06E-03**	**1.02E-02**	**1.12E-01**
MUFA	C15:1n5(cis-10)												4.17E-03				
	C16:1	2.07E-03	1.55E-03				1.45E-03	6.67E-04	2.85E-03				7.83E-03	2.70E-03			8.94E-03
	C18:1n9t				5.17E-03				6.05E-03	5.88E-06			3.94E-03	3.17E-03	5.62E-04	1.09E-03	1.30E-02
	C18:1n9c	1.01E-03		7.74E-05			4.28E-04		8.37E-03				5.52E-03	1.68E-05			1.33E-02
	C20:1n9								1.72E-03				9.39E-04				5.80E-03
	C22:1																2.68E-03
	C24:1n9		2.08E-04		3.56E-04	5.14E-04				1.26E-03				2.90E-03			3.60E-03
	Σ MUFA	**3.08E-03**	**1.76E-03**	**7.74E-05**	**5.53E-03**	**5.14E-04**	**1.88E-03**	**6.67E-04**	**1.90E-02**	**1.27E-03**			**2.24E-02**	**8.79E-03**	**5.62E-04**	**1.09E-03**	**4.73E-02**

		FOZ DO AREIA				SALTO SEGREDO				SALTO SANTIAGO				SALTO CAXIAS			
		ReservoirS															
F. A	**C:D**	**W**	**Sp**	**S**	**Sp2**	**W**	**Sp**	**S**	**Sp2**	**W**	**Sp**	**S**	**Sp2**	**W**	**Sp**	**S**	**Sp2**
PUFA	C18:2n6t								2.09E-03							4.68E-04	
	C18:2n6c				8.06E-04		2.93E-04		2.00E-03								4.69E-03
	C18:3	3.42E-05		9.65E-04							1.69E-04						
	C18:3n3						5.63E-04		2.78E-04				1.10E-03				5.99E-04
	C20:2(cis-11.14)		2.33E-05						9.05E-03	1.46E-03					5.90E-04		
	C22:2																1.41E-04
	Σ PUFA	**3.42E-05**	**2.33E-05**	**9.65E-04**	**8.06E-04**		**8.56E-04**		**1.34E-02**	**1.46E-03**	**1.69E-04**		**1.10E-03**		**5.90E-04**	**4.68E-04**	**5.43E-03**
HUFA	C20:4	1.53E-03					3.49E-04		3.49E-04				1.06E-03				4.74E-04
	EPA	9.09E-03		1.19E-03	1.50E-02	1.38E-04	2.08E-03		2.47E-02	1.17E-04			1.31E-02	9.17E-03	5.89E-04		1.89E-02
	DHA	9.62E-03	8.37E-02	4.01E-05	1.75E-02	2.95E-02	1.06E-03		3.53E-02	8.21E-02		1.88E-02	1.73E-02	1.11E-01		2.01E-02	2.38E-02
	Σ HUFA	**2.02E-02**	**8.37E-02**	**1.23E-03**	**3.25E-02**	**2.96E-02**	**3.49E-03**		**6.03E-02**	**8.22E-02**		**1.88E-02**	**3.15E-02**	**1.20E-01**	**5.89E-04**	**2.01E-02**	**4.32E-02**
	Σ FA	**4.41E-02**	**9.54E-02**	**1.54E-02**	**7.97E-02**	**3.09E-02**	**1.90E-02**	**6.67E-04**	**1.62E-01**	**9.01E-02**	**4.41E-03**	**1.90E-02**	**9.30E-02**	**1.92E-01**	**6.80E-03**	**3.18E-02**	**2.08E-01**
	N	**13**	**7**	**9**	**12**	**5**	**15**	**1**	**16**	**9**	**3**	**3**	**14**	**9**	**6**	**7**	**16**

Table 3.
List of fatty acids identified in the Iguaçu River reservoirs, units in mg of fatty acids per mg of zooplankton (mg.mg^{-1}). W = winter (July 2012), Sp 248 = spring (November 2012), S = summer (February 2013), Sp2 = spring (October 2013). C:D = carbons: saturations.

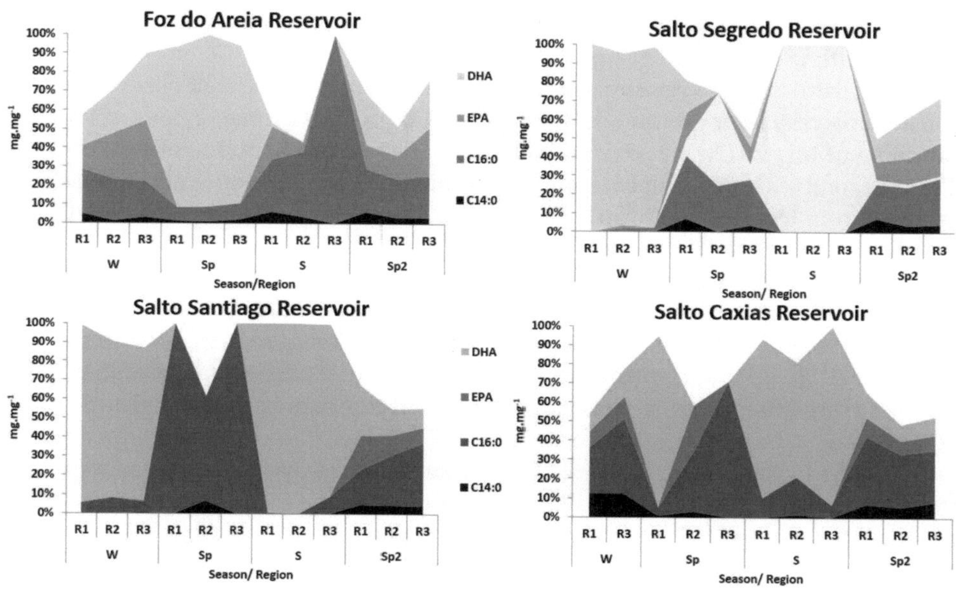

Figure 3.
Fatty acids densities of the Iguaçu River reservoirs.

in these waters. The bloom of the cyanobacterium *Microcystis aeruginosa* is a ubiquitous phenomenon in eutrophic lakes and reservoirs in many countries of the world. According to Cordeiro [28], *Microcystis* has a higher proportion of palmitic acid (C16:0). Adloff [29] in his work with phytoplankton, to the same studied area, found that the four reservoirs have water quality characteristics of eutrophic environments, with intense flowering of cyanobacteria, with a predominance of *M. aeruginosa* and *Sphaerocavum brasiliense*. Flowering occurred mainly in November 2012 and February 2013, and *M. aeruginosa* adapted to the best environment and dominated *S. brasiliense*.

In Salto Caxias reservoir, 19 fatty acids were found, being five saturated fatty acids (SFA), six monounsaturated fatty acids (MUFA), five polyunsaturated fatty acids (PUFA) and three highly unsaturated fatty acids (HUFA) (**Table 3**). In October 2013 sampling (Sp2) was found the most content in fatty acids, 16 in total, of which six are MUFA. In November 2012 sampling (Sp), only 6 fatty acids were identified, of which three were SFA, one PUFA, one MUFA and one HUFA.

In Salto Caxias reservoir, the most abundant fatty acid was DHA in the February 2013 sample (S), being quite representative in region 1 (R1) in November 2012 collection (Sp) with 89% of the total density, and more abundant in the February 2013 sampling (S) contributing more than 93% of all fatty acids (**Figure 3**). As it was not possible to extract the fatty acids from the samples in region 2 (R2) in July 2012 sampling (W), as there was not enough material to extract the fatty acids, only regions 1 (R1) and 3 (R3) were considered.

The collected stations were grouped into regions, with region 1 (R1) covering stations located in the lotic zone of the reservoir—stations P1, P2 and P3; region 2 (R2) includes stations located in the intermediate zone of the reservoir- P4, P5 and P6; and in region 3 (R3) were included the stations in the lake area of the reservoir, those close to the dam—P7, P8 and P9.

In the Cluster analysis, with a cut range of 50%, can clearly be observed the separation of the fatty acid composition according to the season. There a group was formed only with the samples from the collection carried out in spring 2012 (Sp) and the other clusters were composed of samples from other seasons, not being able to differentiate an isolated group.

In the Salto Segredo reservoir, a formation of groups can be well observed in the Cluster analysis. Four groups are distinguished: a group with the regions of the sampling of July 2012 (W), one with all the regions of the sampling of February 2013 (S), one with the samples from November 2012 (Sp) and the other with the samples from October 2013 (Sp2) (**Figure 4**). These results highlight seasonal differences in fatty acid content.

In the Cluster analysis of fatty acids in the Salto Santiago reservoir, it is possible to observe the separation of the fatty acid composition according to the seasons. Groups are formed only with the samples from the collection carried out in the spring of 2013 (Sp2) and spring 2012 (Sp) and the other clusters are composed of samples from other seasons, not being able to differentiate an isolated group (**Figure 4**).

In the Salto Caxias reservoir, the Cluster analysis clearly shows the separation of the fatty acid composition according to the season, where three groups are formed: one with the spring 2012 samples (Sp), another with the samples summer 2013 (S) and another that mostly has samples from spring 2013 (Sp2) (**Figure 4**).

In the multidimensional scaling analysis (n-MDS), in the Foz do Areia reservoir, it was observed the most similar regions according to the composition of fatty acids. Note that the composition of fatty acids was similar when observed the season (Season), showing similarities between the regions (Region), mainly observed in the samples collected in spring of 2012 (Sp). These similarities may indicate a greater homogeneity reservoir, possibly caused by the great drought that occurred there. It was also noted that the region with the greatest dissimilarity was region 3 in February 2013 sampling (S) (**Figure 5**).

In the Salto Segredo reservoir, it is noticed that the composition of fatty acids was similar when observed the season of the year and their respective collection stations, showing similarity mainly in the samples collected in February 2013 (S). This was also observed in the samples of October 2013 (Sp2) and July 2012 (W), with similarities around 60%, reaching up to 80%. In November 2012 (Sp) sampling is observed that the samples from region 2 (R2), show less similarity about the other

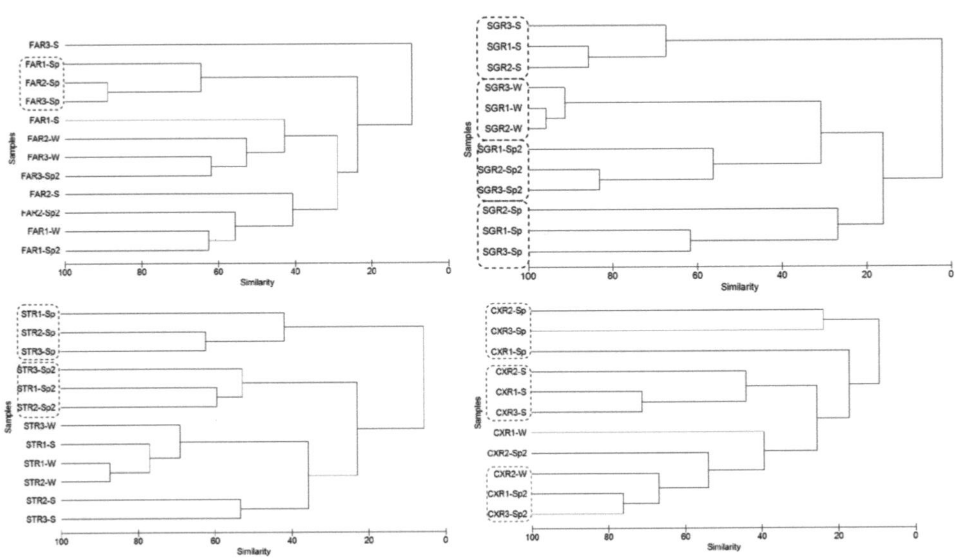

Figure 4.
Cluster analysis of the collection regions according to the seasons in the Iguaçu river reservoirs. W = winter of 2012, Sp = spring 2012, S = summer of 2013, Sp2 = spring 2013.

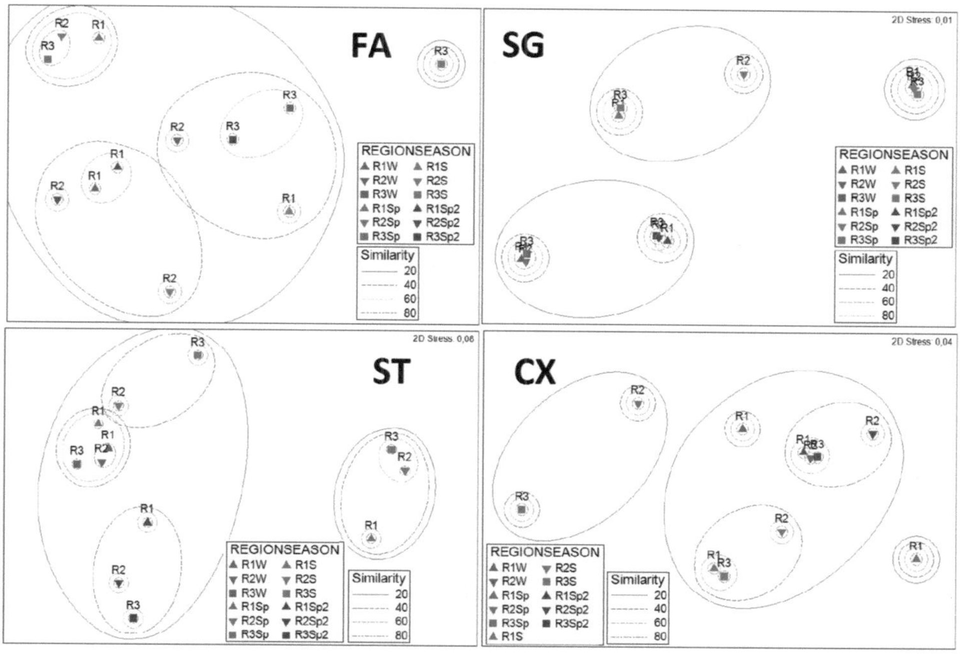

Figure 5.
n-MDS in the Iguaçu river reservoirs according to the fatty acids found. FA = Foz do Areia reservoir, SG = Salto Segredo reservoir, ST = Salto Santiago reservoir, CX = Salto Caxias reservoir.

regions (20%). As this is the intermediate region, that is under the fluvial influences, and also under the lake influences, may showing a different fatty acids content, that may also be related to the composition of the phytoplankton community, from which a large part of the zooplankton community feeds (**Figure 5**). During the sampling in February 2013 (S), in addition to a slight increase in diatom densities [30], there was a great drought in the region, which resulted in a decrease in the level of reservoirs, increasing the concentration of nutrients, which can become an indication of the sudden variation in the concentration of fatty acids.

In the Salto Santiago reservoir, it is clear that the composition of the fatty acids in the Sp and Sp2 samples were similar when observing their respective collected regions (Region), where the similarity is around 60% between the R2 and R3 in the two samples, and less similarity with the R1 region (**Figure 5**).

In the Salto Caxias reservoir, it is noticed that the composition of fatty acids was very different when observed in the collected region (Region). Regions 1 and 3 (R1 and R3) of October 2013 sampling (Sp2) were very similar to region 2 (R2) of the July 2012 sampling (W). Only the February 2013 (S) sample had similar regions (**Figure 5**).

Observing the n-MDS of fatty acids in the reservoirs, it can be noted that in Foz do Areia there is the formation of several groups, where EPA and C16:0 showed a similarity around 60% and DHA a similarity of 40%. The group formed by C14:0 (myristic acid), C16:1 (palmitoleic acid) and C18:0 (stearic acid) shows a similarity of 60% (**Figure 6**).

In Salto Segredo, it is possible to notice the formation of several groups in which EPA and C21:0 (Heneicosylic acid), and DHA and C16:0 (palmitic acid) show a similarity around 60%, one can also perceive a similarity around 80% in the group formed by C18: 1n9c (oleic acid) and C20: 2 (cis-11,14), and also in the group formed by C18: 3n3 (α-linolenic acid) and C20:4 (arachidonic acid).

In the Salto Santiago reservoir, observing the n-MDS of fatty acids, it is possible to notice the formation of two groups, one composed only by C8:0 (caprylic acid)

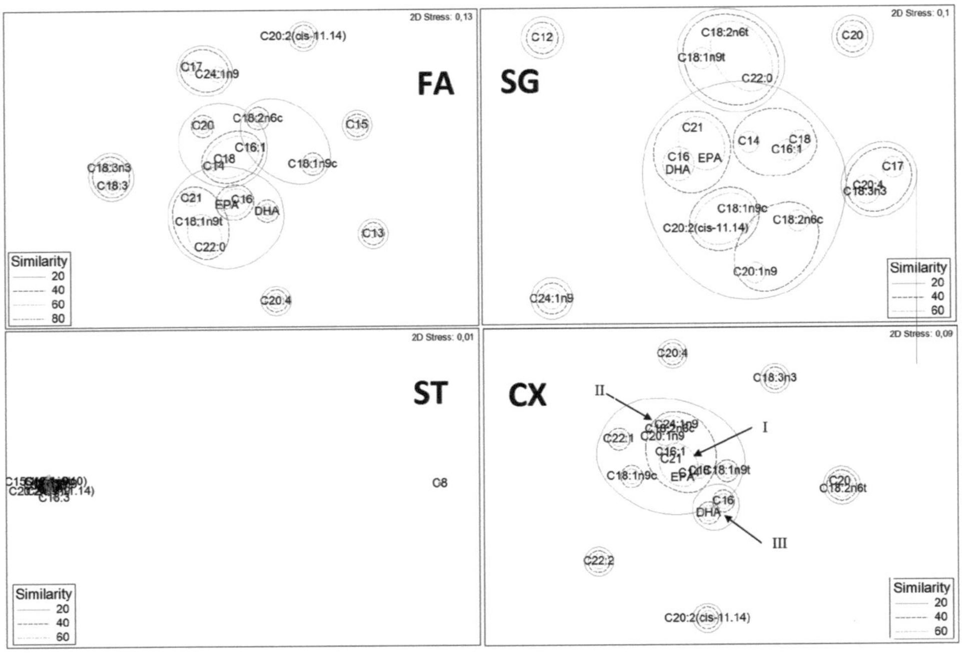

Figure 6.
n-MDS indicating the groups of fatty acids in the Iguaçu river reservoirs. FA = Foz do Areia reservoir, SG = Salto Segredo reservoir, ST = Salto Santiago reservoir, CX = Salto Caxias reservoir.

and the other covers all the other fatty acids in the reservoir. In the Salto Caxias reservoir, it is possible to notice the formation of several groups: group I composed by EPA, C14:0, C21:0 and C16:1, showing a similarity around 60%, as well as group II and group III composed by C20:1n9, C18:2n6, C24:1n9 and DHA, C16:0 respectively (**Figure 6**).

Comparing the fatty acid data with the TSI, it was observed that DHA had a higher concentration in sampling stations where the environment was hypereutrophic, whereas C16:0 was more abundant in sites classified as eutrophic, with other fatty acids like EPA and C20:0 also occurring in some specific sampling sites (**Table 4**).

At the response of zooplankton species and fatty acids behavior about the trophic environmental conditions, a Pearson correlation was made between taxa and fatty acids with the trophic state index (TSI). Regarding zooplankton species, the TSI presented a strong negative correlation with *P. dolichoptera* ($r = -0.8232$; $p = 0.001$), in the Foz do Areia reservoir, which may indicate that this species is more abundant in less eutrophic environments. However, it is clear that, despite the densities of *P. dolichoptera* decrease when the TSI increases, most samples accumulate in TSI values above 58, which means eutrophic environment (**Figure 7**).

In the Salto Segredo reservoir, about taxa, the TSI had a negative correlation with *K. cochlearis* ($r = -0.670$; $p = 0.017$), which may be an indication that this species may be an indicator of an intermediate trophic situation. However, it can be seen that regarding these values, most samples are between 59 and 60, which may mean that this species prefers eutrophic environments (**Figure 7**).

The TSI showed a positive correlation with *P. Dolichoptera* ($r = 0.765$; $p = 0.004$), which may be an indication of the contribution of this rate to the trophic state of the Salto Santiago reservoir. Most of the samples in which *P. dolichoptera* was more abundant were between the values of 55 and 60, showing a preference of these organisms for Mesotrophic to Eutrophic environments,

	TSI × fatty acids															
Collection stations	**FOZ DO AREIA**				**SALTO SEGREDO**				**SALTO SANTIAGO**				**SALTO CAXIAS**			
	W	**Sp**	**s**	**Sp2**	**W**	**Sp**	**s**	**Sp2**	**W**	**Sp**	**s**	**Sp2**	**W**	**Sp**	**s**	**Sp2**
P1	DHA[b]	DHA[f]	C16[d]	C16[d]	X[c]	C16[d]	C16:1[c]	C16[c]	C8[c]	C16[c]	X[b]	C16[c]	DHA[c]	DHA[c]	DHA[d]	C16[b]
P2	C16[b]	DHA[f]	C20[d]	C16[d]	DHA[c]	C16[d]	C16:1[d]	X[d]	DHA[c]	C16[e]	C16[c]	DHA[d]	C16[c]	C16[c]	C16[e]	C16[c]
P3	EPA[a]	DHA[f]	C16[e]	C16[d]	X[c]	C16[d]	C16:1[d]	C16[d]	DHA[c]	C16[f]	DHA[c]	C16[e]	C16[c]	C16[c]	DHA[c]	C16[a]
P4	DHA[a]	DHA[f]	C16[d]	C16[d]	X[c]	C16[d]	X[d]	DHA[c]	DHA[c]	X[c]	DHA[a]	C16[c]	C16[c]	C16[c]	X[c]	C16[b]
P5	EPA[b]	DHA[f]	C20[d]	C16[d]	X[c]	C16:1[d]	C16:1[d]	DHA[d]	DHA[c]	C16[d]	X[c]	C16[d]	C16[c]	C16[c]	C16[d]	C16[c]
P6	DHA[b]	DHA[f]	C16[d]	C16[d]	DHA[c]	C16[d]	C16:1[d]	X[c]	C16[c]	C16[d]	DHA[c]	C15:1n5	X[c]	C16[c]	DHA[d]	C16[c]
P7	DHA[b]	DHA[f]	C16[c]	DHA[d]	X[d]	C16[d]	C16:1[c]	X[d]	DHA[c]	C16[d]	DHA[c]	C15:1n5[d]	X[d]	C16[c]	X[c]	C16[c]
P8	C16[b]	DHA[f]	C16[d]	DHA[d]	C16[d]	C16[d]	C16:1[c]	C16[d]	DHA[c]	C16[c]	C16[b]	C16[c]	X[c]	C16[c]	C16[d]	C16[b]
P9	X[b]	DHA[f]	C16[d]	DHA[d]	DHA[d]	C18:1n9[d]	X[d]	X[c]	DHA[c]	X[c]	DHA[d]	C15:1n5[d]	X[c]	C16[c]	DHA[f]	C16[d]

[a]*Ultraoligotrophic*
[b]*Oligotrophic*
[c]*Mesotrophic*
[d]*Eutrophic*
[e]*Supereutrophic*
[f]*Hipereutrophic*

Table 4.
Trophic state index (TSI) compared to the most abundant fatty acid species in the four Iguaçu River reservoirs.

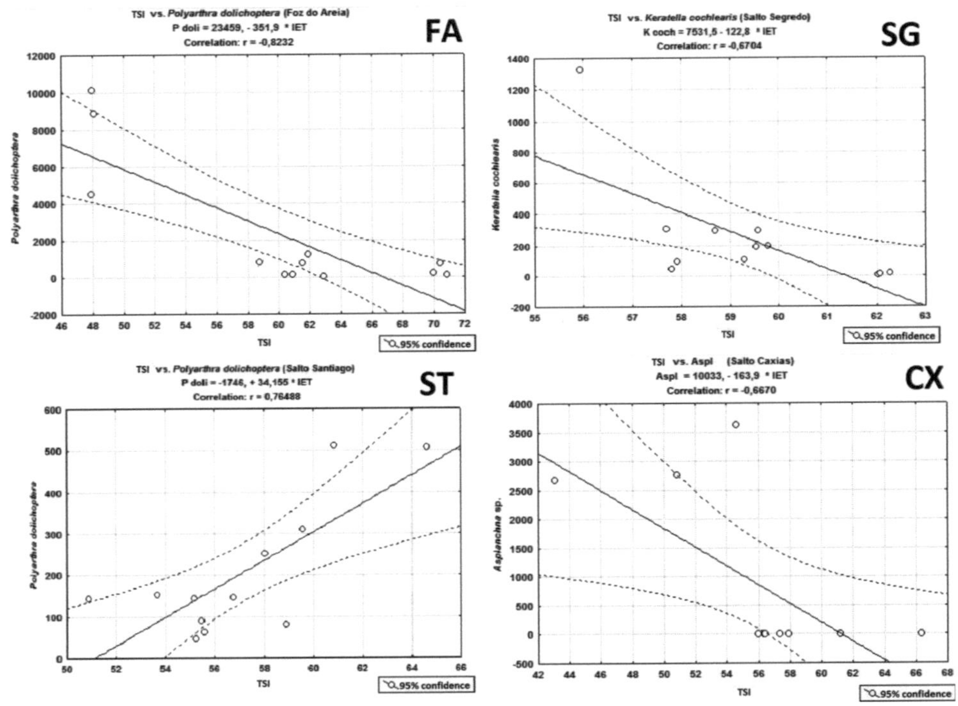

Figure 7.
Pearson correlation between TSI and most abundant zooplankton taxa in Iguaçu river reservoirs. B) Pearson correlation between TSI and DHA in the Foz do Areia reservoir.

although the highest density was found in the supereutophic range (63 < TSI <67) (**Figure 7**).

In the Salto Caxias reservoir, the TSI had a strong negative correlation with *Asplanchna* sp. ($r = -0.667; p = 0.025$) (**Figure 7**), which may be indicative of improved water quality over time, as this genus is a predator, and usually appears when there is a lot of food available, like herbivorous rotifers, and when food is scarce. Consequently, there is an improvement in the trophic state, since the photosynthetic organisms were consumed by the herbivores and there is a large amount of these small herbivores left, which are preyed upon by *Asplanchna* sp., and at this stage, the environment is already recovering from the large nutrient load caused by the activities changes and phytoplankton senescence.

Regarding fatty acids, the TSI had a strong negative correlation with oleic acid (C18:1n9c) ($r = -0.7392; p = 0.006$) and a strong positive correlation with docosahexaenoic acid (DHA) ($r = 0.666; p = 0.018$) (**Figure 8**), in the Foz do Areia reservoir. When the environment was hypereutrophic (TSI > 67), the highest concentrations of DHA appeared, however, most samples were in the range between 59 to 63, related to the eutrophic environment, although the concentrations were not very high.

In the Salto Segredo reservoir, regarding fatty acids, the TSI had a strong negative correlation with DHA ($r = -0.648; p = 0.023$), which may indicate that this fatty acid has its highest concentrations in less eutrophic environments. However, it can be seen that there are more samples of fatty acids in more eutrophic environments, however, with lower concentrations (**Figure 8**).

In the Salto Caxias reservoir, the TSI had a strong negative correlation with fatty acids C16:0 ($r = -0.610; p = 0.046$), C18:3n3 ($\mathrm{r} = -0.748; p = 0.008$), C20:1n9 ($\mathrm{r} = -0.663; p = 0.026$) and EPA ($\mathrm{r} = -0.611; p = 0.046$) (**Figure 8**), which may

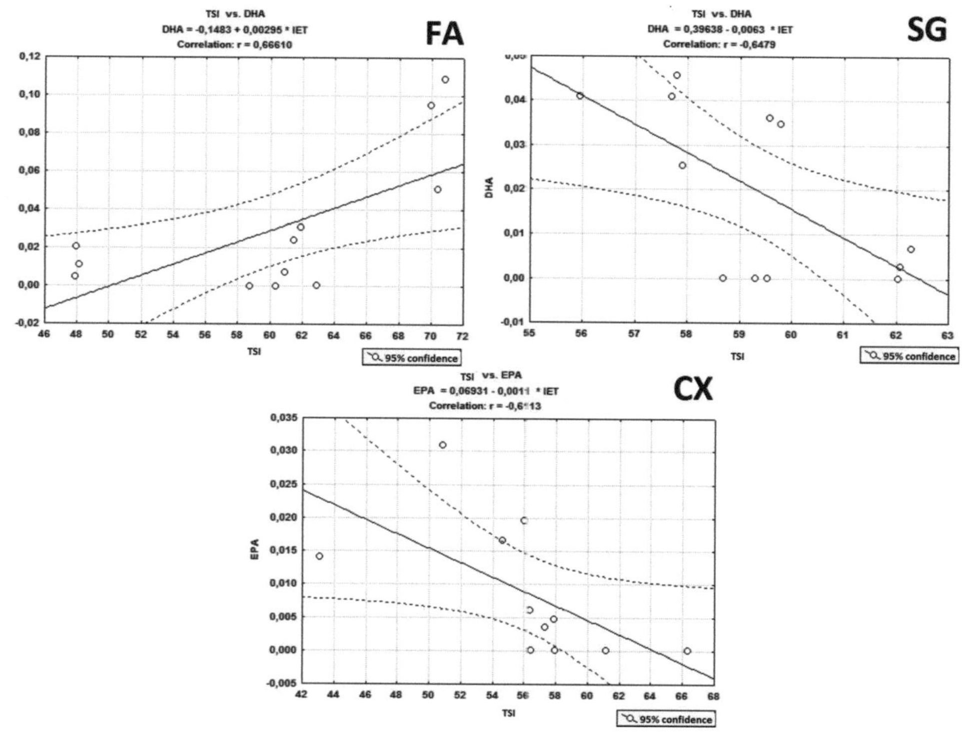

Figure 8.
Pearson correlation between TSI and most important fatty acids in the Iguaçu river reservoirs.

indicate that the reduction in water quality in relation to the TSI may result in a decrease in the concentrations of fatty acids above. In the Salto Santiago reservoir, the TSI did not show any significant correlation with fatty acids, which may be an indication that these fatty acids are not related to the trophic state of the reservoir.

The principal component analysis (PCA) between zooplankton species and fatty acids in Foz do Areia explained 75.45% of the total variability, with the first main component (PC1) explaining 44.22% and the second main component (PC2) 31.23%. PC1 (Factor 1) is separated into two distinct groups between the most abundant species and fatty acids. PC1 positively explained the following variables: Eicosapentaenoic acid (EPA) (0.96), *Docosahexaenoic* acid (DHA) (0.84), *B. longirostris* (0.82), as well as a large group of fatty acids. PC1 negatively indicated an association between DHA, α and γ-linolenic acids (C18:3n3 and C18:3, respectively) and the species *Synchaeta jollyae* (Synch jo) and *K. cochlearis* (K coch), which may indicate that these fatty acids can be associated with these species (**Figure 9**), since many organisms can synthesize DHA from α-linolenic acid (18,3n3) found in algae and plants.

Observing PC2 (Factor 2), it was noticed that there was a positive separation, mainly between *B. longirostris*, myristic acid (C14:0) and elaidic acid (C18:1n9t). That may indicate that this cladoceran species may be related to these fatty acids, as *Bosmina* species is a highly selective consumer [31], not absorbing some forms of fatty acids, preferentially feeding on EPA-producing algae. On the negative side of PC2, it was noticed that there is a great relationship between the copepod nauplii (Náup), *C. cornuta* (Cd corn) and *P. dolichoptera* (P doli) with oleic acid (C18:1n9c), which can be indicative that this fatty acid was important for the densities of these taxa (**Figure 9**). It has been suggested that myristic (C14:0), palmitic (16,0), and oleic (C18:1n9c) acids are derived from fatty acids from algae [32, 33].

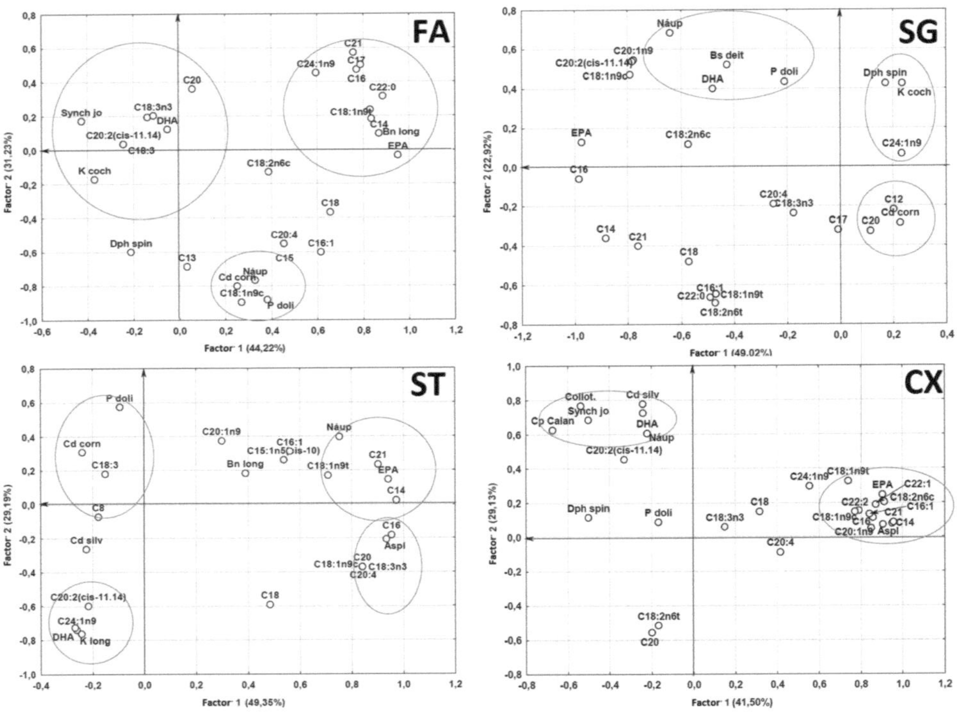

Figure 9.
Association between fatty acids and the most abundant taxa in the Iguaçu River reservoirs during the study period.

In the Salto Segredo reservoir, the principal component analysis (PCA) among the fatty acid data with the most abundant taxa explained 67% of the total variability, with the first main component (PC1) explaining 49.02% and the second main component (PC2) 22.92%. The first main component (PC1) separated two groups of distinct variables.

The positive side of the first component explained better the relationship between the following variables: the taxa *C. cornuta* (Cd corn), *K. cochlearis* (K coch) and *D. spinulosum* (Dph spin) and the fatty acids, lauric acid (C12: 0), arachidic acid (C20: 0) and nervous acid (C24: 1n9). These associations may indicate that these species may be feeding on phytoplankton that have these fatty acids. The negative side of the first component shows the association between DHA, *B. deitersi*, *P. dolichoptera* and copepod nauplii (Náup). These associations may indicate that certain environmental conditions provide the production of DHA and the zooplankton community is receiving these fatty acids through food. The second main component located on the positive side, associated taxa with some types of fatty acids indicating that these fatty acids may be contributing to the development of species. On the negative side, there is only a grouping of fatty acids (**Figure 9**).

The principal components analysis between fatty acids and the most abundant taxa in the Salto Santiago reservoir explained 78.54% of the total variability, with the first main component (PC1) explaining 49.35% and the second main component (PC2) 29.19%, and two distinct groups were formed, being explained by the first main component (PC1). The larger group located on the positive side indicates that the genus *Asplanchna* sp., influenced several fatty acids, possibly contributing to the increase of its densities, such as C16:0, C18:3n3, C18:1n9c, C20:0 and C20:4.

The variables contained in the other ellipse show a greater association between copepod nauplii and C14:0, C21:0 and EPA fatty acids, indicating that the concentrations of these fatty acids may have influenced nauplii densities. The second group formed by the ellipse on the negative side of factor 1 showed that the concentrations of DHA, C24:1n9, C20:2(cis-11,14), may have influenced the abundances of the rotifer *Kelicottia longispina* (K long), which was the most abundant rotifer in this reservoir. The second main component (factor 2) explained the relationship between *P.dolichoptera* and *C.cornuta* with C18:3 (**Figure 9**).

In the Salto Caxias reservoir, the principal component analysis, between fatty acids and taxa, explained 70.63% of the total variability, with the first main component (PC1) explaining 41.50% and the second main component (PC2) 29.13%. The positive side of factor 1 explained the relationship of *Asplanchna* sp. With several fatty acids, mainly with EPA, C14:0 and C16:0. These associations may indicate that a certain group of algae may be producing these fatty acids, which favored the abundance of rotifers, which are the food source of *Asplanchna* sp. The second main component, located on the positive side, associated DHA with several taxa, among them *S. jollyae*, *Collotheca* sp., *C. cornuta*, copepods and calanoid copepodites, indicating that these taxa may possibly be related to a diet rich in DHA (**Figure 9**).

4. Discussion

This study highlights the predominance of zooplankton species that are somehow adapted to eutrophic environments, corroborating, in part, with the observed trophic condition of the reservoir, with the aquatic system classified as oligotrophic, according to the TSI in winter of 2012 (W). In the four reservoirs, the dominant species were similar. *P. dolichoptera* was the most abundant species among the rotifers and *C. cornuta,* whereas in the cladoceran species *Ceriodaphnia silvestrii, D. spinulosum* and *B. longirostris* were the most abundant. In November 2012 (Sp) and February 2013 (S) collections, it may have a large phytoplankton density, and herbivorous rotifers such as *P. dolichoptera* become dominant over other species, and later can be controlled by small zooplanktonic carnivorous organisms [34], which may explain the dominance of *Asplanchna* sp. In upstream reservoirs, namely, Salto Santiago and Salto Caxias.

According to Hollowday [35], *P. dolichoptera* is found mainly in eutrophic environments, but also occur in water bodies with different trophic state degrees, but present higher densities in eutrophic water bodies in periods of low temperature [36]. This dominance of *P. dolichoptera* may indicates that environmental eutrophication could occur, which was confirmed in the Foz do Areia reservoir in the spring of 2012 (Sp). Regarding the TSI, the trophic environmental conditions were quite different, because in the spring of 2012 (Sp) the region was experiencing a great drought, and the reservoirs were below their normal level and present a probable "bloom" of cyanobacteria *M. aeruginosa*, according to Adloff *et al*. [29].

Due to this combination of factors, probably resulted in changes in trophic status in aquatic bodies. Another factor that may have been the cause of this eutrophic condition, mainly in the Foz do Areia reservoir, is the dumping of waste into the Iguaçu River, in the region of the city of Curitiba, as it is in a region with little access to basic sanitation, as well as, the presence of industries.

Over time, these wastes accumulate in the first reservoir, and combined with the drought, the high temperature and the low level of the reservoir, increases the concentration of nutrients, consequently making it hypereutrophic. This high concentration of nutrients in the Foz do Areia reservoir passes to the other reservoirs along the river, a fact that was possible to see in the following samples, arriving at the Salto Caxias Reservoir, in the last collection. In the last sampling, the reservoir floodgates were opened, which may have caused these nutrients to be carried along the river, however, as this study was carried out during this process, it is not possible to infer the impact caused.

Farage et al. [37] observed an opposite result, in which the trophic state of the studied aquatic environment increased during the rainy season, which is justified by the runoff mechanism, which often occurs during the rainy season, especially in soils without vegetation or predominance undergrowth.

Observing the results of the analysis of fatty acids superimposed with the TSI, it is noticed that the main compounds are DHA, C16:0. In lesser proportions occurred EPA and C15:1n5. Thant appears in certain trophic states, as in the spring sampling of 2012 (Sp), where in the Foz do Areia reservoir it was presented as hypereutrophic at all sampling sites, where the most abundant fatty acid was DHA. However, DHA was more abundant in places classified as mesotrophic in the 2012 winter sampling (W) in the Salto Santiago reservoir.

In the Salto Caxias reservoir, environments classified as mesotrophic, C16:0 fatty acid was the most abundant. This variation may have occurred due to changes in the climate and the water regime of the reservoirs, in the spring of 2012 (Sp), where there was a great drought, and the reservoir levels decreased considerably.

This variation may have occurred due to changes in the climate and the water regime of the reservoirs, in the spring of 2012 (Sp), where there was a great drought, and the reservoir levels decreased considerably. In Foz do Areia the environment was classified as hypereutrophic, and DHA was the most abundant fatty acid while the other fatty acids had a lower concentration. A dominance of DHA may indicate a diet based on flagellate organisms [38]. The seasonal variation of fatty acids, mainly DHA, indicates a temporal change in the availability of phytoplankton [39]. Observing the results obtained, DHA is more present in more eutrophic environments, where according to Nozaki [40], some flagellated algae are more abundant, while EPA, as well as C14:0 and C16:0 are more abundant in less eutrophic environments.

Palmitic acid (C16:0) is a fatty acid that is found in several groups of phytoplankton that are food sources for zooplankton [41] and also in some species of Cyanophyceae [42]. Also, according to Patil et al. [42], C16:0 is one of the main groups of saturated fatty acids (SFA) among freshwater phytoplankton, while marine phytoplankton is the main producer of DHA, and little is found in freshwater species, considering that DHA is the most important fatty acid for copepods and many fish species [43, 44]. Indeed, palmitic acid is one of the most common saturated fatty acids in organisms, in general. Generally, phytoplankton with high proportions of EPA or DHA, such as Cryptophyceae and Bacillariophyceae are excellent food sources for zooplankton. Cyanophytes have practically no EPA and DHA, while diatoms are rich in EPA, and dinoflagellates have high amounts of DHA [45].

In aquatic ecosystems, the level of essential fatty acids (EFAs), such as EPA and DHA, in algae is highly variable [46]. The HUFA content represents between 3 and 7% of the total fatty acids of phytoplankton during flowering, making the nutritional value of phytoplankton flowering questionable [46]. There is also evidence that the amount of EPA and DHA in algae varies significantly between the major taxonomic groups [47]. For example, Cryptophyceae have high proportions of EPA and DHA, while in Chlorophyceae it is nonexistent, scarce or has traces of these fatty acids. There is ample evidence that essential fatty acids (EFA) such as eicosapentaenoic acid (EPA) and docosahexaenoic acid (DHA) are extremely important for several aquatic organisms [48]. Indeed, marine phytoplankton generally contains high amount of omega-3 (n-3) fatty acids, which are good for human health being one of the main foods sources the consumption in fish species. These essential fatty acids are related with the prevention and treatment of ocular, cardiovascular, autoimmune and cancer diseases [48–50].

The availability of PUFAs is a critical factor that influences trophic dynamics and biological production in the ecosystem. However, this PUFA pool is in turn influenced by the set of planktonic organisms present, as well as physical and chemical factors. To date, few studies have analyzed the transfer efficiencies of these fatty acids at various trophic levels.

The concentration of zooplanktonic fatty acids is a useful tool for defining the trophic state of the environment, especially EPA, DHA and C16:0. It is known that the zooplankton community is important for the flow of energy in the trophic chain. According to the concentration of nutrients in the environment, there may be a differentiation in the biochemical composition of this community. To understand how these differences can be related to water quality is an issue to be further studied.

Although, zooplankton is not included as BQE (Biological Quality Element) in the WFD (Water Frame Directive of the European Union Water Framework), some studies use this community to assess the water quality of aquatic systems [51, 52]. This work corroborates zooplankton is a good bioindicator to assess water quality and thus identified as a BQE by Water Framework Directive.

Biochemical analyzes of zooplankton, such as lipids (including fatty acids), showed to be an important tool for water body management and water quality analysis, and for detecting molecular changes in the zooplankton community, related to the trophic status of systems, before they are perceived in the environment.

Although, the use of fatty acid analysis is not a low-cost tool and can only be used with the help of a specialist, the results obtained are more accurate than other types of analysis, thus being able to generate faster conclusions, accelerating the assessment and evaluation process of the aquatic bodies.

Public policies and environmental campaigns should be adopted by the responsible environmental agencies to minimize the discharge of sewage into the river, in the region of Curitiba, or if possible, its treatment, so that this load of pollutants does not reach the first reservoir—Foz do Areia—as well as environmental education campaigns for the population.

Acknowledgements

Fabiano Ramiro Serpe thanks to CAPES (CAPES/PDSE 99999.014500/2013-02) for the financial support during the exchange period in Portugal and the granted PhD scholarships. This work was financed by national funds through FCT (Foundation for Science and Technology), I.P., within the scope of the projects UIDB/04292/2020 (MARE, Marine and Environmental Sciences Centre) and UIDP/50017/2020 + UIDB/50017/2020 (CESAM, Centre for Environmental and Marine Studies). The project PORBIOTA - E-Infrastructure Portuguese Information and Research in Biodiversity (POCI-01-0145-FEDER-022127) co-financed this research, supported by Competitiveness and Internationalization Operational Programme and Regional Operational Programme of Lisbon, through FEDER, and by the Portuguese Foundation for Science and Technology (FCT), through national funds (OE). Ana M. M. Gonçalves acknowledges the University of Coimbra for the contract IT057-18-7253.

Author details

Fabiano Ramiro Serpe[1], Júlio César Rodrigues de Azevedo[1,2], Fernando J.M. Gonçalves[3], João C. Marques[4] and Ana M.M. Gonçalves[3,4]*

1 PPGERHA—Postgraduate Program in Water Resources and Environmental Engineering, Federal University of Paraná (UFPR), Curitiba, Brazil

2 DAQBI—Academic Department of Chemistry and Biology, Federal Technological University of Paraná (UTFPR), Curitiba, Brazil

3 Department of Biology and CESAM—Centre for Environmental and Marine Studies, University of Aveiro, Aveiro, Portugal

4 Department of Life Sciences, Calçada Martim de Freitas, University of Coimbra, MARE—Marine and Environmental Sciences Centre, Coimbra, Portugal

*Address all correspondence to: amgoncalves@uc.pt

References

[1] Bashir I, Lone FA, Bhat RA, Mir SA, Dar ZA, Dar SA. Concerns and threats of contamination on aquatic ecosystems. In: Hakeem K, Bhat R, Qadri H, editors. Bioremediation and Biotechnology. Cham: Springer; 2020

[2] Cairns J, McCormick PV, P. V. and B. R. Niederlehner, B. R. A proposed framework for developing indicators of ecosystem health. Hydrobiologia. 1993; **263**:1-44

[3] Lomartire S, Marques JC, Gonçalves AMM. Biomarkers based tools to assess environmental and chemical stressors in aquatic systems. Ecological Indicators. 2021;**122**:1-9

[4] Arnott SE, Celis-Salgado MP, Valleau RE, DeSellas AM, Paterson AM, Yan ND, et al. Road salt impacts freshwater zooplankton at concentrations below current water quality guidelines. Environmental Science & Technology. 2020;**54**(15):9398-9407

[5] Li Y, Chen F. Are zooplankton useful indicators of water quality in subtropical lakes with high human impacts? Ecological Indicators. 2020;**113**:106-167

[6] Wang L, Chen J, Haojie S, Xufa M, Zhixu W, Hong S, et al. Zooplankton Body Size an Indicator of Water Quality in (Sub)tropical Reservoirs in China? Ecosystems. London, UK: Springer Nature; 2021. pp. 1-13

[7] Moss B. Ecology of Fresh Waters: A View for the Twenty-First Century. 4th ed. Oxford, UK: John Wiley & Sons; 2010. 470 p

[8] Hausmann S, Charles DF, Gerritsen J, Belton TJ. A diatom-based biological condition gradient (BCG) approach for assessing impairment and developing nutrient criteria for streams. Science of the Total Environment. 2016;**562**: 914-927

[9] Poikane S, Zampoukas N, Borja A, Davies SP, van de Bund W, Birk S. Intercalibration of aquatic ecological assessment methods in the European Union: Lessons learned and way forward. Environmental Science & Policy. 2014;**44**:237-246

[10] Poikane S, Birk S, Böhmer J, Carvalho L, de Hoyos C, Gassner H, et al. A hitchhiker's guide to European lake ecological assessment and intercalibration. Ecological Indicators. 2015;**52**:533-544

[11] Jeppesen E, Nõges P, Davidson TA, Haberman J, Nõges T, Blank K, et al. Zooplankton as indicators in lakes: A scientific-based plea for including zooplankton in the ecological quality assessment of lakes according to the European water framework directive (WFD). Hydrobiologia. 2011;**676**:279-297

[12] Gonçalves AMM, Azeiteiro UM, Pardal MA, de Troch M. Fatty acid profiling reveals seasonal and spatial shifts in zooplankton diet in a temperate estuary. Estuarine, Coastal and Shelf Science. 2012;**109**:70-80

[13] Shurin JB, Havel JE, Leibold MA, Pinnel-Alloul B. Local and regional zooplankton species richness: A scale-independent test for saturation. Ecology. 2000;**81**:3062-3073

[14] Filimonova V, Gonçalves F, Marques JC, De Troch M. Fatty acid profiling as bioindicator of chemical stress in marine organisms: A review. Ecological Indicators. 2016;**67**:657-672

[15] Gonçalves AMM, Rocha CP, Marques JC, Gonçalves F. Fatty acids as suitable biomarkers to assess pesticide impacts in freshwater biological scales – A review. Ecological Indicators. 2021; **122**:107299

[16] Perbiche-Neves G, Serafim-Júnior M, Portinho JL, Shimabukuro EM, Ghidini AR, de Brito L. Effect of atypical rainfall on lotic zooplankton: Comparing downsteam of a reservoir and tributaries with free stretches. Tropical Ecology. 2012;**53**(2):149-162

[17] Eletrosul. O impacto ambiental da ação do homem sobre a natureza – rio Iguaçu, Paraná, Brasil – Reconhecimento da ictiofauna, modificações ambientais e usos múltiplos de reservatórios. Florianópolis/SC: Eletrosul; 1978. 33 p

[18] Takeda AM, Butakka CMM, Fujita DS, Bibian JPR. Larvas Chironomidae em Cascata de Reservatorios no rio Iguaçu, Paraná. In: Rodrigues L, Thomaz SM, Agostinho AA, Gomes LC, editors. Biocenoses em Reservátorios: padrões espaciais e temporais. São Carlos/SP: RIMA; 2005. pp. 147-160

[19] da Silva PRL. Efeitos de Reservatórios em Cascata sobre a Biologia Reprodutividade um Piscívoro Neotropical. Toledo/PR. Dissertação de mestrado: UNIOESTE; 2010. 50 p

[20] Julio-Junior HF, Thomaz SM, Agostinho AA, Latini JD. Distribuição e caracterização dos reservatórios. Biocenoses em reservatórios: padrões espaciais e temporais. São Carlos: Rima; 2005. pp. 1-16

[21] Gubiani EA, Frana VA, Maciel AL, Baumgartner D. Occurrence of non-native fish Salminus brasiliensis (Cuvier, 1816) in a global biodiversity ecoregion, Iguaçu River, Paraná River Basin, Brazil. Aquatic Invasions. 2010;**5** (2):223-227

[22] Mendonça F. Riscos, vulnerabilidade e abordagem socioambiental urbana: uma reflexão a partir da RMC e de Curitiba. Desenvolvimento e Meio Ambiente. 2004;**10**:139-148

[23] Haney JF, Hall DJ. Sugar-coated daphnia : A preservation technique for Cladocera. Limnology and Oceanography. 1973;**18**:331-333

[24] APHA—American Public Health Association. Standard Methods for the Examination of Water and Wastewater. 20th ed. New York: American Public Health Association; 2005

[25] Lamparelli MC. Graus de trofia em corpos d\'água do estado de São Paulo: avaliação dos métodos de monitoramento [doctoral dissertation]. Universidade de São Paulo: São Paulo; 2004

[26] Berberovic R, Pinto-Coelho R. Dry first, measure later: a new procedure to preserve and measure zooplankton for ecophysiological studies. Journal of Plankton Research. 1989;**11**(5): 1109-1116

[27] Pedersen L, Jensen HM, Burmeister AD. The significance of food web structure for the condition and tracer lipid content of juvenile snail fish (Pisces: Liparis spp.) along 65–72°N off West Greenland. Journal of Plankton Research. 1999;**21**:1593-1611

[28] Cordeiro R. Cianobactérias isoladas de lagos naturais do Parque Estadual do Rio Doce-MG potencialmente produtoras de biocombustível. Belo Horizonte/MG: UFMG; 2015

[29] Adloff CT, Bem CC, Reichert G, Azevedo JCRD. Análise da comunidade fitoplanctônica com ênfase em cianobactérias em quatro reservatórios em sistema de cascata do rio Iguaçu, Paraná. Brasil: RBRH; 2018. p. 23

[30] Serpe CTA. Análise da composição e abundância da comunidade fitoplantônica em quatro reservatórios do rio Iguaçu. In: Universidade Tecnológica Federal do Paraná (UTFPR). Curitiba-PR: Dissertação; 2014. 180 pp

[31] DeMott WR, Kerfoot WC. Competition among cladocerans: Nature of the interaction between Bosmina and daphnia. Ecology. 1982;**63**:1949-1966

[32] Ahlgren G, Gustafsson IB, Boberg M. Fatty acid content and chemical composition of freshwater microalgae. Journal of Phycology. 1992;**28**:37-50

[33] Desvilettes C, Bourdier G, Amblard C, Barth B. Use of fatty acids for the assessment of zooplankton grazing on bacteria, protozoans and microalgae. Freshwater Biology. 1997;**38**:629-637

[34] Smith VH, Crews T. Applying ecological principles of crop cultivation in large-scale algal biomass production. Algae Research. 2014;**04**(4):23-34

[35] Hollowday ED. Family Synchaetidae. In: Nogrady T, Segers HR, editors. Guides to the Identification of the Microinvertebrates of the Continental Waters of the World. Vol. 18. The Hague, the Netherlands: SPB Academic Publications; 2002. pp. 87-211

[36] Negreiros NF, Santos RM, Oliveira LLD, Santos-Wisniewski MJ, Rocha O. Synchaeta jollyae (Shiel & Koste) (Rotifera, Synchaetidae): na overlooked species or a newcomer in Neotropical freshwaters? Iheringia, Série Zoologia. 2012;**102**(2):159-162

[37] Farage JAP, Matos AT, Silva DD, Borges AC. Determinação do índice de estado trófico para fósforo em pontos do rio pomba. Engenharia na agricultura. 2010;**18**(4):322-329

[38] Dalsgaard J, St John M, Kattner G, Müller-Navarra D, Hagen W. Fatty acid trophic markers in the pelagic marine environment. Advances in Marine Biology. 2003;**46**:225-340

[39] Couturier LIE, Rohner CA, Richardson AJ, Marshall AD, Jaine FRA, Bennett MB, et al. Stable isotope and signature fatty acid analyses suggest reef manta rays feed on demersal zooplankton. PLoS One. 2013;**8**(10):1-15

[40] Nozaki H. Flagellated green algae. In: Wehr JD, Sheath RG, editors. Freshwater Algae of North America. Amsterdam, Netherlands: Elsevier Inc; 2003. p. 918

[41] Taipale S, Strandberg U, Peltomaa E, Galloway AWE, Ojala A, Brett MT. Fatty acid composition as biomarkers of freshwater microalgae: Analysis of 37 strains of microalgae in 22 genera and in seven classes. Aquatic Microbial Ecology. 2013;**71**:165-178

[42] Patil V, Källqvist T, Olsen E, Vogt G, Gislerød HR. Fatty acid composition of 12 microalgae for possible use in aquaculture feed. Aquaculture International. 2007;**15**:1-9

[43] Sargent JR, Bell G, McEvoy GL, Tocher LD, Estevez DA. Recent developments in the essential fatty acid nutrition of fish. Aquaculture. 1999;**177**: 191-199

[44] Watanabe T. Importance of docosahexaenoic acid in marine larval fish. Journal of the World Aquaculture Society. 1993;**24**:152-161

[45] Brett MT, Müller-Navarra DC. The role of highly unsaturated fatty acids in aquatic food-web processes. Freshwater Biology. 1997;**38**:483-499

[46] Morris RJ, McCartney MJ, Robinson GA. Studies of a spring phytoplankton bloom in an enclosed experimental ecosystem. I. Biochemical changes in relation to the nutrient chemistry of water. Journal of Experimental Marine Biology and Ecology. 1983;**70**:249-262

[47] Sargent JR, Whittle KJ. Lipids and hydrocarbons in the marine food web. In: Longhurst AR, editor. Analysis of Marine Ecosystems. London: Academic Press; 1981. pp. 491-533

[48] Arts MT, Ackman RG, Holub BJ. "Essential fatty acids" in aquatic ecosystems: A crucial link between diet and human health and evolution. Canadian Journal of Fisheries and Aquatic Sciences. 2001;**58**(1):122-137

[49] Holub BJ. Dietary fish oils containing eicosapentaenoic acid and the prevention of atherosclerosis and thrombosis. Canadian Medical Association Journal. 1988;**139**:377-381

[50] Leaf A. Cardiovascular effects of fish oils: Beyond the platelet. Circulation. 1990;**82**:624-628

[51] Kiggundu V, Mwebaza-Ndawula L, Makanga B, Nachuha S. Variations in zooplankton community structure and water quality conditions in three habitat types in northern Lake Victoria. Lakes & Reservoirs: Research and Management. 2012;**17**(2):83-95

[52] Ren L, Zhang Z, Zeng X, Ma Y, Zeng Y, Zhou C. Community structure of zooplankton and water quality assessment of Jialing River in Nan Chong. Procedia Environmental Sciences. 2011;**10**:1321-1326